Infrastructure Procurement and Funding

Infrastructure is vital to a resilient society and infrastructure investment is therefore critical to the vibrant functioning of societies. Infrastructure assets span economic and social spheres, but despite the prime importance of infrastructure investment, national governments simply cannot fund all of society's infrastructure requirements. This book, *Infrastructure Procurement and Funding* explores the key models of procuring and financing major projects and infrastructure works whilst critically acknowledging the inherent challenges in successfully securing the necessary funding.

The book provides the reader with a detailed review of contemporary methods of financing and procuring infrastructure projects, commencing with an examination of the role of infrastructure in society in creating resilient societies. It reviews public sector funding mechanisms for infrastructure investment before presenting emerging trends in private sector investment in infrastructure. Fundamentally this book identifies robust, innovative and contemporary solutions to the procurement, financing and investment in major infrastructure projects, globally, nationally and regionally.

The book is ideal reading for international courses in construction procurement, construction project management, infrastructure asset management, real estate investment and finance, but will also be useful for those construction business leaders in public and private sectors who are responsible for making major project and infrastructure financial and investment decisions.

Sharon McClements PhD. MRICS SFHEA, PGCHEP, BSc Hons is a chartered quantity surveyor, Senior fellow of the Higher Education Academy and Chair of the RICS Northern Ireland. She lectures at Ulster University in a range of Built Environment programmes and supervises PhD researchers in areas of infrastructure and BIM. She has presented her research at international conferences and has secured funding grants for her research and consultancy. Sharon won the Herbert Walton Excellence Award, from APM, for her PhD thesis. Sharon is also an Expert Author on the Infrastructure Framework and Risks for RICS.

Infrastructure Procurement and Funding

Harnessing Investment to Deliver a Better Future

Sharon McClements

LONDON AND NEW YORK

Cover image: © Frank Lee/Getty Images

First published 2022
by Routledge
4 Park Square, Milton Park, Abingdon, Oxon OX14 4RN

and by Routledge
605 Third Avenue, New York, NY 10158

Routledge is an imprint of the Taylor & Francis Group, an informa business

British Library Cataloguing-in-Publication Data
A catalogue record for this book is available from the British Library

Library of Congress Cataloging-in-Publication Data
Names: McClements, Sharon, author.
Title: Infrastructure procurement and funding : harnessing investment to deliver a better future / Sharon McClements.
Description: Milton Park, Abingdon, Oxon ; New York, NY : Routledge, 2022. | Includes bibliographical references and index.
Identifiers: LCCN 2022003470 (print) | LCCN 2022003471 (ebook) | ISBN 9780367775278 (hardback) | ISBN 9780367767525 (paperback) | ISBN 9781003171805 (ebook)
Subjects: LCSH: Infrastructure (Economics) | Public works—Finance.
Classification: LCC HC79.C3 M43 2022 (print) | LCC HC79.C3 (ebook) | DDC 363—dc23/eng/20220126
LC record available at https://lccn.loc.gov/2022003470
LC ebook record available at https://lccn.loc.gov/2022003471

ISBN: 978-0-367-77527-8 (hbk)
ISBN: 978-0-367-76752-5 (pbk)
ISBN: 978-1-003-17180-5 (ebk)

DOI: 10.1201/9781003171805

Typeset in Goudy
by codeMantra

Contents

Preface

Brief description of this book

Globally there is a relentless pressure on investing more in urban infrastructure, caused by the compounding effect of i) continuously increasing demands upon ageing and decaying assets; and ii) the ever-escalating expectation for modern, sustainable and more innovative infrastructure systems. These factors have created a funding crisis within governments, ultimately exacerbating the urban infrastructure investment deficit. Furthermore, governments have sustained cuts to overall investment in infrastructure due to the financial impact of the global financial crisis of 2008.

Based on my extensive review of infrastructure investment, it has become apparent that the approaches used by governments to harness greater investment in infrastructure are changing– in particular changes to both the funding streams and the procurement strategies adopted. My research has identified that these changes, in both funding and procurement of infrastructure, are due to the devolvement of powers to the regions and the increasing complexities required to fund modern societies. This book explores the contemporary range of financial models and procurement strategies that can deliver greater investment in infrastructure.

Furthermore, it demonstrates contemporary challenges and innovative solutions to funding and procuring infrastructure works. In doing so this book identifies synergies between funding needs and funding streams and demonstrates that effective investment in infrastructure can act as a 'sinew', connecting places, people and businesses, on a local and international level, and with the essential services they need.

The need for this book

The aim of this book is to provide a contemporary review of infrastructure funding models and procurement strategies, making it essential reading for infrastructure practitioners requiring greater understanding of the challenges and opportunities in investing in infrastructure; for policy makers seeking to align infrastructure policy implementation with best practice; and for academics and students who seek a greater understanding of the complexities of investing in modern societies through greater investment in infrastructure.

Key features of this book

This book is set out in a logical manner and begins with the exploration of the challenges to greater infrastructure investment. It considers the different infrastructure demands placed upon Municipals, developed nations and developing nations. In addition, I have included case studies to demonstrate the effective and efficient delivery of infrastructure projects around the world. Finally, the book considers the emerging challenges and opportunities to greater infrastructure investment.

I hope you enjoy it.

1 Introduction

Investment of infrastructure should be at the heart of Government

Primary aim of this book

Infrastructure investment is critical to the vibrant functioning of society. Ergo every citizen is impacted by the investment made into infrastructure projects. Despite the prime importance of infrastructure investment, globally governments cannot fund all of society's infrastructure requirements. This book explores the importance of infrastructure to our daily lives and the investment strategies adopted by governments to ensure high quality infrastructure provision. Furthermore, this book presents the key models of procuring and financing major projects and infrastructure works whilst critically acknowledging the inherent challenges in successfully sustaining the required levels of infrastructure investment. This volume provides readers with a contemporary review of innovative infrastructure investment solutions and frameworks adopted by Government to level up the infrastructure investment deficit in the UK.

The structure of this chapter

This chapter considers the role of infrastructure in creating and maintaining a vibrant, healthy and sustainable society for all citizens. Commencing with an overview of the meaning and classification of infrastructure as either an economic or social asset, it further analyses the trends in national infrastructure investment, depicting the multi-faceted nature of the widening infrastructure investment deficit, referred to widely as the infrastructure investment 'gap'. Finally, it explores Government's ambitious national and regional investment plans aimed at transforming the infrastructure assets and delivering economic recovery across the four regions of the UK.

This chapter has been structured as follows:

- Definition of infrastructure – What is infrastructure?
- The role of infrastructure in promoting a sustainable society
- Infrastructure investment and the investment gap
- National infrastructure investment plans
- Regional infrastructure investment
- Summary

DOI: 10.1201/9781003171805-1

Learning outcomes

1. This chapter presents an overview of the strategic importance of infrastructure investment. Globally, infrastructure investment is considered to be one of the most important drivers of productivity and growth in national economies.

    - The economic importance of infrastructure investment stems from the investment potential that drives and stimulates economic activities nationally and locally. The stimulation of economic activity creates a vibrant market that maximises innovation in ideas, technologies and processes.

2. Contemporary analysis is given concerning capital infrastructure investment whilst acknowledging the key factors that have created an infrastructure investment deficit or 'gap'.

    - This chapter articulates the consequences of underinvestment in infrastructure and identifies these consequences or negative impacts on society.

3. Finally, it identifies the ambitions of the UK Government to increase infrastructure investment and investment opportunities so as to reconcile the negative impacts on society of long-term sustained underinvestment in infrastructure.

Definition of infrastructure – What is infrastructure?

Infrastructure is vital for the development and functioning of society. The term infrastructure emanated out of World War II where it was initially employed as a military reference to denote 'underlying' structures. Subsequently, 'infrastructure' was adopted by development economists to describe 'social overhead capital' and is now a widely recognised expression (Howes and Robinson, 2005). Yet, as there is growing demand for greater investment in national infrastructure projects, the term infrastructure has now become increasingly attractive to both Government and investors, even though, at the same time, it has also been widely acknowledged there has been, globally, a long-term sustained lack of investment. Therefore, 'infrastructure' now has political, economic and environmental attributes.

Historically, infrastructure has been characterised as capital intensive projects and services which are funded by taxpayers or Government borrowing, and regarded as essential public services. Despite the public service essence of infrastructure there has been a long tradition of private sector collaboration in the provision of infrastructure projects. This collaboration has resulted in public–private partnership (PPP) models which, over the last decade, have matured and developed, and this has, over time, led to the acceptance by society that many of these 'public' services can be provided, maintained and funded by private investors. Nonetheless infrastructure is still considered to be an essential public service. Grimsey and Lewis (2002) agree, adding that 'infrastructure comprehensively includes public services, the economic sector as well as social contributors which influence living standards and quality of life'.

Over the years there have been many attempts to categorise the term infrastructure and to define the concept of an essential public service. For example, Rutherford (2002) defined infrastructure as 'the basic services or social capital of a country, or a part of it, which make economics and social activities possible', where as Adetola et al. (2011) argued that rather than the traditional focus on physical 'hard' assets such as roads, ports, communications, energy and water services, infrastructure should now be considered as an umbrella term additionally accompanying 'softer' services such as information technology and knowledge bases.

Notwithstanding these contributions, increasingly the expression infrastructure has been classified as economic or social infrastructure. Conventionally, economic infrastructure is considered to be hard or physical structures/assets such as roads, bridges, railways, ports and public buildings that are essential for a functioning society and which support local and national economic development. Support of economic development is derived through the flow of people and goods and also the delivery of public services such as highways, ports, airports, utilities, power and communication systems. These are typically long-standing, capital intensive, engineered network structures built to bolster economic activity. Comparatively, social infrastructure is considered public service or buildings such as municipal, leisure, education, emergency services, health, housing, justice and recreational assets which are fundamental to the development of society by enhancing the quality of life and living standards for all (Howes and Robinson, 2005). This book, therefore, will present and discuss infrastructure in the context of both economic and social infrastructure.

For further clarification, this volume adopts the classification of economic and social infrastructure as given Figure 1.1

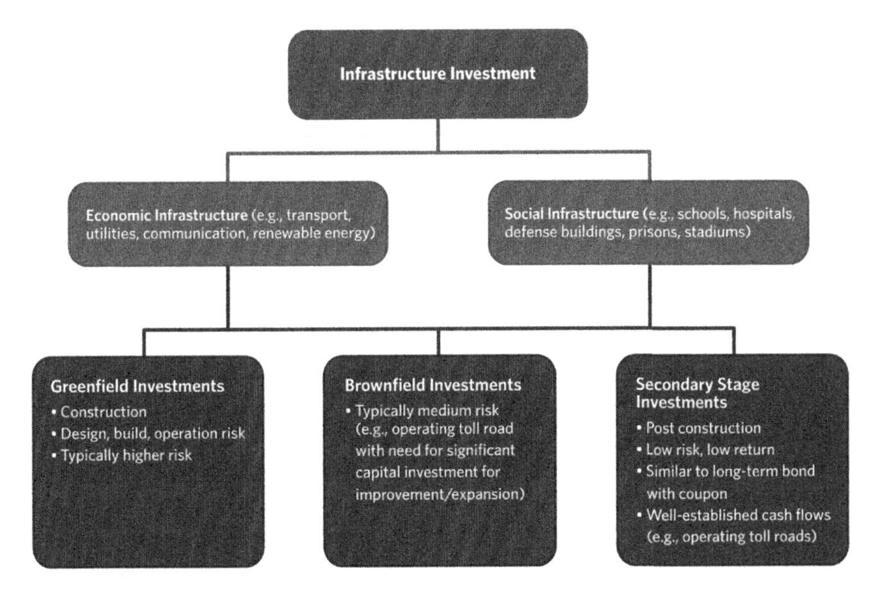

Figure 1.1 Types of infrastructure investment
Source: World Bank (2012).

The role of infrastructure in promoting a sustainable society

Investment in public infrastructure, to deliver an essential public service or maintain a public asset, is of national importance. Infrastructure investment is considered important as it has been shown to have a direct economic contribution nationally and regionally and can deliver a range of social benefits to local communities. Significantly, these local economic and community benefits contribute nationally through job growth and creation, and to the overall well-being and health of citizens, creating vibrant, attractive and sustainable communities. Yet, despite the national importance of infrastructure investment in creating such communities, governments cannot entirely fund, invest and manage these infrastructure needs. Instead, they must determine and plan infrastructure needs and seek to procure and finance infrastructure using innovative and collaborative investment models.

The Covid-19 pandemic has created havoc across the globe; no country is immune, no sector within society is immune. The brunt of shielding and treating those afflicted with the virus lies with the health and social services sector. In the UK, the Governments have borrowed record amounts, not seen since World War II, in an attempt to mitigate the social and economic disaster of this devasting virus. Billions of pounds have been pumped into the NHS; unemployment has soared, as businesses have been forced to close or at best operate on a much lower scale; and individuals have found themselves furloughed under the Government's financial support scheme for the unemployed and those put on unpaid leave. In addition to tax breaks, rates relief and business grants, the UK Government has borrowed £355bn in the financial year ending April 2021 (OBR, 2021) to tackle the Covid-19 pandemic.

Whilst short-term funding and financial support packages are undoubtedly necessary, economists are now pondering the long-term economic impact of the pandemic or, more specifically, the impact of borrowing to tackle this pandemic. It was originally purported that the recession caused by Covid-19 would be short and sharp and the recovery would also rebound sharply. The 'V'-shaped scenario initially suggested by economists seems to have faded. Although the pandemic may not cause a great depression as witnessed after World War I, it will have a long-lasting impact that will result in a slower and more gradual 'U'-shaped recovery. Yet, a recovery, nonetheless. So, the question to leaders seeking to learn lessons from this pandemic must be how to make society and the economy resilient and deliver sustainable healthy communities.

One way to achieve this is to invest in infrastructure. Infrastructure is vital to a resilient society and its assets span both the economic and social spheres. There are a plethora of data showing the UK's sustained underinvestment in infrastructure and it has been well-documented. This book therefore offers a timely examination of funding and procurement models which seek to close the gap in this underinvestment, giving the perceptions of key stakeholders concerning alternative infrastructure procurement and funding mechanisms to create resilient and healthy communities.

Infrastructure investment is critical to the vibrant functioning of society. Ergo every citizen is impacted by the investment made in infrastructure projects. Despite the prime importance of infrastructure investment, as highlighted earlier, governments cannot fund all of society's infrastructure requirements that are essential for resilient and healthy communities. This book explores this increasing dilemma of needing to invest more in infrastructure whilst acknowledging that governments do not have the funds to procure and finance infrastructure projects. Acknowledging this dilemma, it provides a timely and contemporary review of the key models of procuring and financing major projects and infrastructure works whilst critically acknowledging the inherent challenges in successfully securing the necessary funding. In doing so this volume provides readers with a contemporary review of the main procurement and financing models that have been developed in recent times to mitigate the investment gap.

The UK's infrastructure requires a radical upgrade in order to keep pace with society's demands. Despite Government funding of infrastructure, it has become apparent and been acknowledged that the private sector will be required to fill the growing deficit in infrastructure finance and investment. Raising finance privately via either equity or debt is now considered critical in making infrastructure projects a reality.

Hence, this book will examine the strategic importance of infrastructure investment in terms of creating and maintaining sustainable and vibrant growth nationally and regionally. Central to this examination is the identification and appraisal of both public and private infrastructure finance and investment mechanisms that deliver sustainable infrastructure investment globally, nationally and regionally. Thus, this book is relevant for construction business leaders, in both public and private sectors, who are responsible for making financial and investment decisions on major projects and infrastructure.

Fundamentally, investment in infrastructure should be at the heart of Government. Investment in infrastructure can take us to a better place economically, socially and environmentally. Modern and sustained infrastructure is a key building block for prosperity.

Infrastructure investment

Delivering sustainable infrastructure requires substantial investments. The Paris Climate Change Accord, the UN Sustainable Development Goals, the Habitat III New Urban Agenda and the Sendai Framework for Disaster Risk Reduction have all shown the need for a more strategic approach to investing in public infrastructure; one that both channels and utilises private and public capital investment more effectively. Therefore, increasingly it has been acknowledged that a broad and diverse class of investors are required to deliver the capital investment and long-term funding for infrastructure projects both creatively and collaboratively. Combining national strategic needs with local community needs must also be considered by governments pursuing sustainable infrastructure growth and investment and for tackling the investment gap.

Infrastructure investment gap

According to the OECD (2020), globally, at current investment trends there will be a cumulative investment gap of between 'US$5.2 trillion until 2030 or as high as US$14.9 trillion until 2040 when the achievement of the sustainable development goals (SDGs) is taken into account'. Previous research conducted by McKinsey (2016) had already recognised that the global infrastructure investment gap was increasing. In fact, according to a range of estimates, the deficit in investment for global infrastructure is growing by more than US$1 trillion annually, as can be seen from the OECD (2020) figures.

This investment gap is particularly acute in developing countries and emerging economies, highlighting the underlying economic differences between developing and developed countries. However, because the economic, social and political divergences between developed and developing countries is too great to reconcile in one book, we will here consider the infrastructure funding and procurement models of developed countries only.

Whilst governments are deliberating on the complexities of funding and procuring infrastructure, they are also required to deliver infrastructure investment sustainably. The World Economic Forum (2013) has estimated that an additional US$0.7 trillion per year would be needed to move from the 'business-as-usual' economy to green growth. Others have suggested that sustainable infrastructure can often carry higher capital costs and technical risks.

The current global pandemic has exposed chronic underinvestment in infrastructure globally. If this infrastructure gap is not closed, every citizen in society will face difficulties pertaining to sanitation, clean drinking water, energy provision, waste collection, adequate shelter, public transport systems and general accessibility. Ultimately, people's mobility for employment, livelihood and quality of life will be inhibited. Oxford Economics (2017) has modelled historic, current and future infrastructure investment need (see Figure 1.2).

Figure 1.2 clearly depicts the infrastructure investment gap globally by identifying historic, current and future investment needs. Fundamentally the model

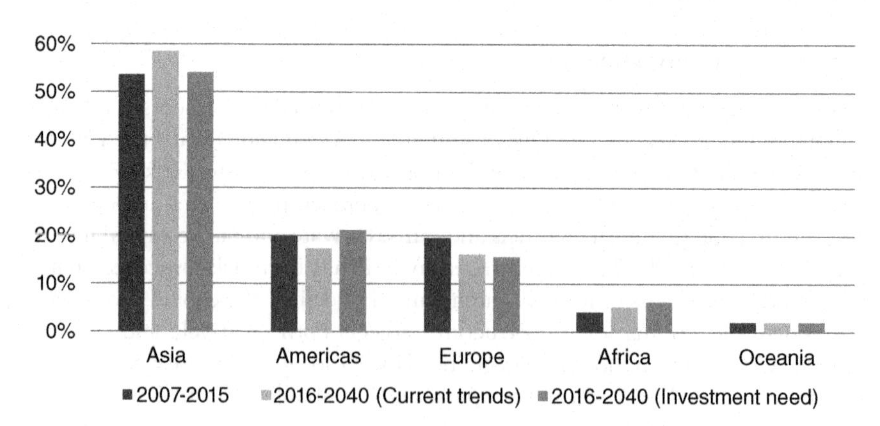

Figure 1.2 Regional share of global infrastructure investment, 2007–2040
Source: Oxford Economics (2017).

predicts that a sustained and widening infrastructure investment gap is unavoidable if governments continue to commit to just the current levels of infrastructure investment. As can be seen in more detail in Figure 1.3, the Oxford Economics model suggests the Americas have the greatest need for investment (47%), followed by Africa (39%), Europe (16%), Oceania (10%) and Asia (10%). Combined, there is a need for an increase in infrastructure investment from US$2,300bn to US$3,300bn from 2015 to 2030.

However, as the Oxford Economics report states, 'Changes in the structure of global infrastructure spending are more apparent when data are viewed in terms of geography. Infrastructure investment in Asia increased by more than 50 percent

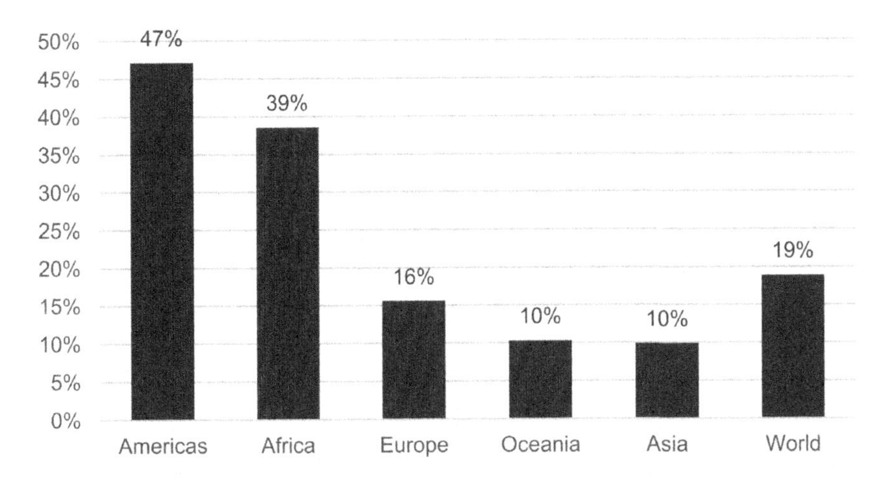

Figure 1.3 Infrastructure investment gap by region, 2016–2040. Extent to which estimated investment need is greater than investment expected under current trends
Source: Oxford Economics (2017).

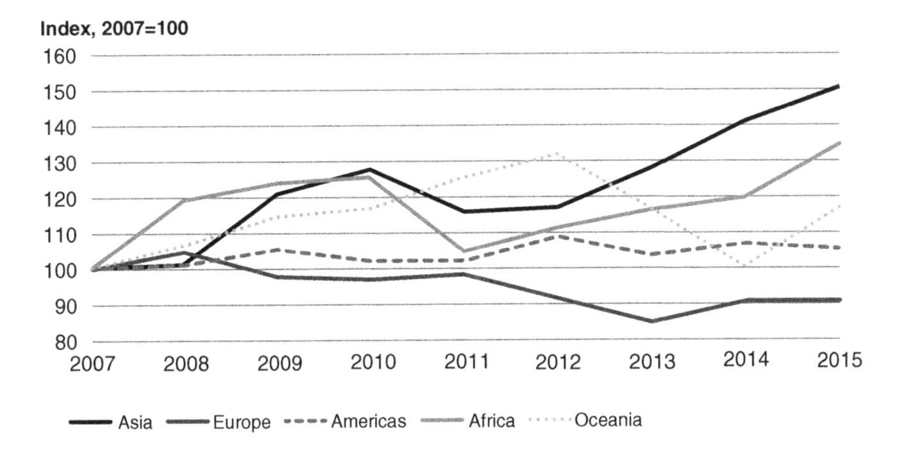

Figure 1.4 Regional infrastructure investment, 2007–2015.
Source: Oxford Economics (2017).

between 2007 and 2015 [whereas] spending in Europe fell back …, partly in response to the constrained state of government finances' (see Figure 1.4).

UK infrastructure investment

International infrastructure rankings provide a way of comparing the performance of UK infrastructure with other countries. According to the World Economic Forum, in 2019 the UK was ranked 11th out 141 countries in terms of the overall quality of its infrastructure, behind France (7th), Germany (8th), and the Netherlands (2nd). The USA was ranked 13th.

Also, the Organisation for Economic Cooperation and Development's (OECD, 2015)report found that 'gross government investment … [was] lower as a share of GDP in the UK than in other countries'. According to ONS (2018), the majority of infrastructure funding comes from the public sector, either centrally or locally. As an example, the UK invested almost £19bn in infrastructure in 2016, with £12bn coming from central government and £7bn from local government (see Figure 1.5).

More recent analysis produced by the OBR (2021) shows that, since the 1970s, the Government has actually been committed to spending more on infrastructure (see Figure 1.6). Historically, trends in public sector net investment showed that net capital investment was at a modest 1.6 per cent of GDP in March 2020, which was much lower than the average investment of 4 per cent of GDP. This stagnation in capital investment in infrastructure impacted the UK's ranking of infrastructure quality. As mentioned above, the World Economic Forum (2019) ranked the UK 11th for infrastructure quality, behind both France and Germany. According to the TUC's (2021) ranking of all G7 countries' green recovery and jobs investments, the UK comes 6th. Only Japan scores worse per person.

The low levels of public sector investment, combined with the soaring public sector debt of £2.3 billion (96.1 per cent of GDP), will necessitate new methods of financing infrastructure (ONS, 2021). Therefore, significant levels of investment will be required in UK infrastructure in the coming years to meet the UK Government's objectives for economic growth and de carbonisation. Towards this

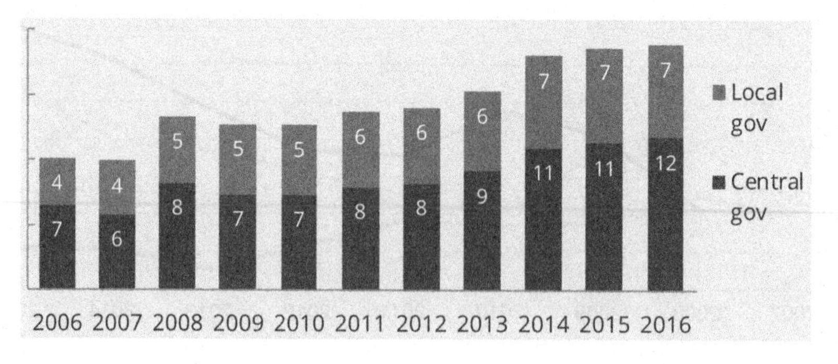

Figure 1.5 Public sector investment in infrastructure (£ billions, UK)
Source: ONS (2018).

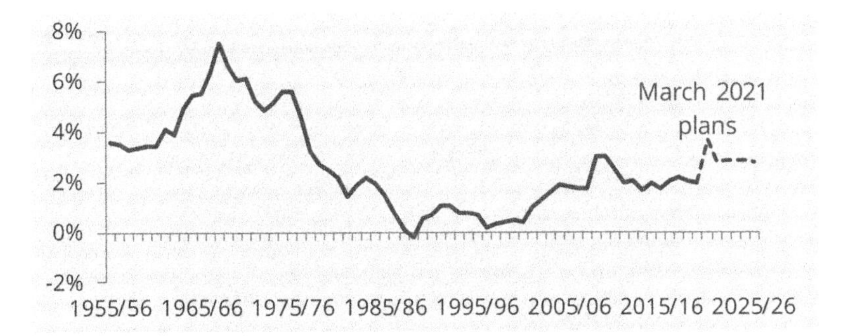

Figure 1.6 Government plans public sector net investment of 3 per cent of GDP
Source: OBR (2021, March).

goal, the UK Spending Review 2020 (HM Treasury, 2020) committed £100 billion of capital investment in 2021–22;an almost £30 billion cash increase compared with 2019–20. These plans make progress on delivering the UK Government's objective of over £600 billion of investment over the next five years, representing the highest sustained levels of investment as a proportion of GDP since the late 1970s. Additionally, the UK Government has announced a series of infrastructure investment reviews which have resulted in the publication of both national and regional investment plans. The ultimate purpose of these investment plans is to provide clarity and perhaps certainty to investors of the UK Government's commitment to reduce the infrastructure investment deficit.

National infrastructure investment plans

Prior to the pandemic, the UK was emerging from a decade of austerity. The UK, like most modern developed economies, entered a long period of austerity measures designed to repay the economic damage caused by the global financial crisis in 2008. Emerging almost a decade later, the UK Conservative Government commissioned an assessment into the infrastructure needs of the nation. The National Infrastructure Commission (NIC) was tasked with completing this assessment. This assessment resulted in the Government announcing infrastructure investment plans around transport, energy and water networks.

Despite the NIC assessment of infrastructure needs, and Government rhetoric on vital and strategic investment into transport, energy and water, the Institute of Civil Engineers (Goodwin,2020) argued that

> there is less concrete progress in the strategy on the frameworks and governance arrangements that will inevitably be required for regions (specifically across England) to better plan and deliver sub-national infrastructure provision. This is key if the government's 'levelling-up' agenda, for which it has announced a new £4bn fund, is to come to fruition.

Figure 1.7 Pipeline of infrastructure investment in the UK for 2020/21
Source: Infrastructure and Projects Authority (2020).

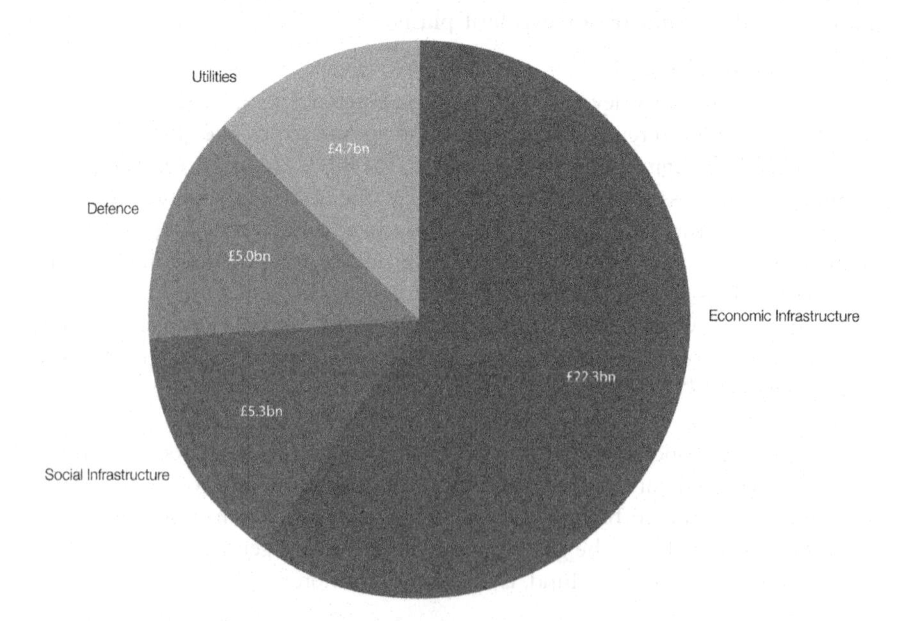

Figure 1.8 Estimate maximum contract value of procurement in the pipeline by sector (£bn)
Source: Infrastructure and Projects Authority (2020).

Thus, ICE was signalling that UK infrastructure was stagnating further and that the stagnated investment would have a negative impact on the economy and the direct impact on communities would also increase.

A further review on the Government's infrastructure investment commitments was conducted in June 2020 by the UK Infrastructure and Projects Authority (IPA). The IPA review, Analysis of the National infrastructure and Construction Procurement Pipeline 2020/21, presented an infrastructure investment pipeline for over 260 infrastructure projects, including 340 procurement contracts, programmes and other investments. The IPA analysis established that the infrastructure procurement pipeline amounted to an estimated contract value of up to £37 billion over the next year (IPA, 2020; see Figure 1.7).

The IPA analysis further highlighted the key infrastructure sectors of major development, as can be seen in Figure 1.8.

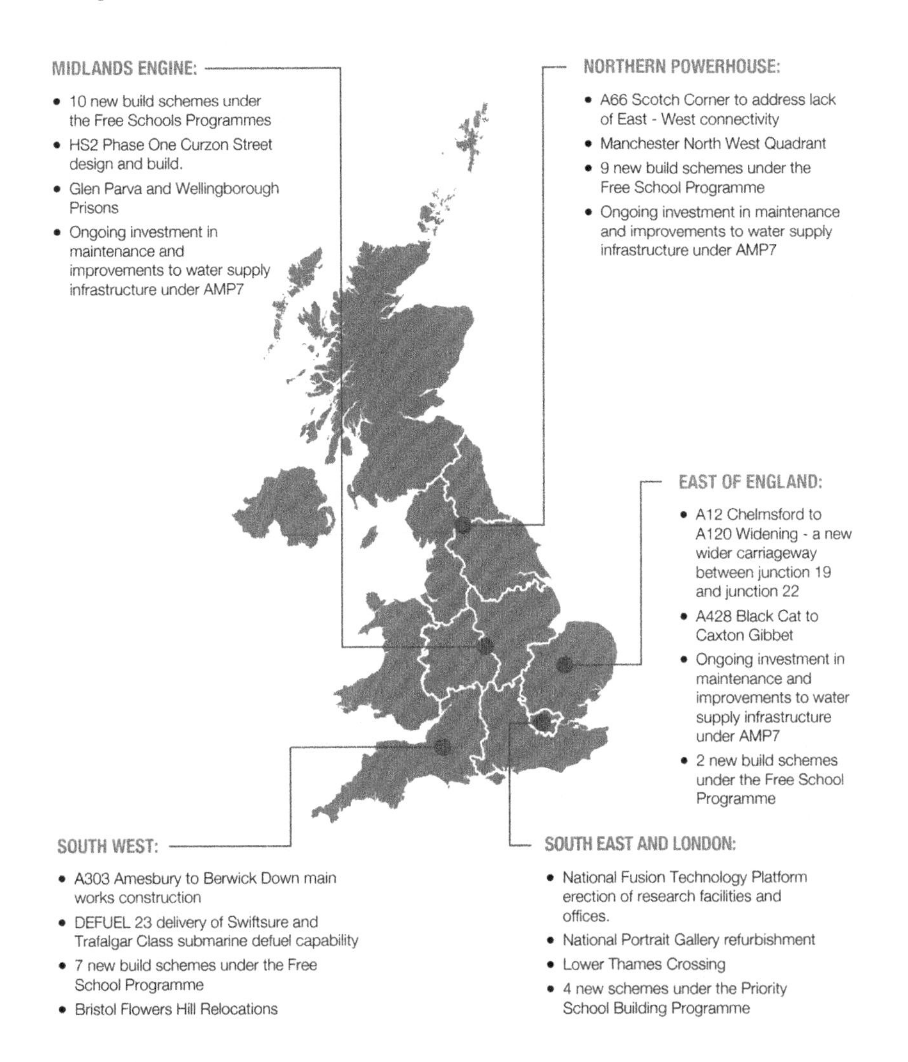

Figure 1.9 Infrastructure project pipeline, England
Source: Infrastructure and Projects Authority (2020).

Figure 1.8 identifies the estimated value of infrastructure contracts as classified into the key sectors of economic, social, defence and utilities. Of the £37.3bn contract, 60 per cent of the infrastructure contracts procured are considered economic infrastructure, 14 per cent are social infrastructure projects, 13 per cent are defence and 13 per cent are classified as utilities infrastructure. This highlights that the Government's focus is on economic infrastructure projects such as transportation and renewable energy projects.

Regionally, strategically important infrastructure projects were also identified in the Government's procurement pipeline. Figure 1.9 highlights the specific investment in various 'powerhouse' locations in England (Infrastructure and Projects Authority, 2020). (Note that in the UK, Northern Ireland, Scotland and Wales are devolved regions and have devolved responsibility for planning and prioritising their regional infrastructure requirements.)

UK *National Infrastructure Strategy*

Following on from the infrastructure pipeline publication, the UK Government presented a National Infrastructure Strategy in November 2020. The strategy, aimed at delivering the identified £37.3bn infrastructure investment, was subtitled 'Fairer, Faster, Greener'. This was to reflect the Government's desires to invest in infrastructure that would benefit all in society whilst meeting the demands of a carbon-neutral society. Specific to the 'fairer' agenda, the strategy acknowledges the need for a 'levelling up' approach to infrastructure investment. The levelling up mantra identifies Government's acceptance that the infrastructure gap needs to be addressed. Therefore, the key commitments of the Government's 'Fairer, Faster, Greener' strategy highlights the following infrastructure investment initiatives:

- The Government is setting up a new UK Infrastructure Bank to co-invest alongside the private sector in infrastructure projects.
- The bank will operate UK-wide, be based in the North of England, and support the Government's ambitions on levelling up and net zero
- The bank will also be able to lend to local and mayoral authorities for key infrastructure projects and provide them with advice on developing and financing infrastructure.
- The Government is committed to the model of independent economic regulation but will refine it to ensure it provides a clear and enduring framework for investors and businesses and delivers the major investment needed in decades to come, while continuing to deliver fair outcomes for consumers.
- The Government will produce an overarching policy paper on economic regulation in 2021, which will consider regulator duties, how to inject more competition into strategic investments and the benefits of a cross-sectoral Strategic Policy Statement; and
- The Government will continue to develop new revenue support models and consider how existing models – such as the Regulated Asset Base model and

> Contracts for Difference – can be applied in new areas and remains open to new ideas from the market. The Government will not reintroduce the private finance initiative model (PFI/PF2).

In addition to investing more in infrastructure projects, in March 2021 the Chancellor Rishi Sunak presented the Government's spring budget, which set out the Government's 'build back better' strategy for economic growth. This strategy concluded that

> high quality infrastructure is crucial for economic growth, boosting productivity and competitiveness. Infrastructure helps connect people to each other, people to businesses, and businesses to markets, forming a foundation for economic activity and community prosperity.

Furthermore, the Chancellor stated that this high-quality infrastructure should be at the centre of our communities. To build back better, in his budget the Chancellor identified the following goals in infrastructure investment:

> Stimulate short-term economic activity and drive long-term productivity improvements via record investment in broadband, roads, rail and cities, as part of … capital spending plans worth £100 billion next year.
>
> Connect people to opportunity via the UK-wide Levelling Up Fund and UK Shared Prosperity Fund, as well as the Towns Fund and High Street Fund, to invest in local areas.
>
> Help achieve net zero via £12 billion of funding for projects through the Ten Point Plan for a Green Industrial Revolution.

Regional (devolved) infrastructure investment plans

In the UK, HM Treasury holds overall responsibility for setting infrastructure spend for England; however, this responsibility is devolved in Scotland, Wales and Northern Ireland. To assist the devolved regions in determining infrastructure investment decisions HM Treasury publishes key policy, guidance and statistics on infrastructure procurement and funding and offers advice to all departments undertaking or wishing to undertake infrastructure projects. By providing standard guidance and policy on infrastructure investment the Treasury aims to safeguard that any private investment into public infrastructure assets delivers value for money, supports the regional and national economy and delivers targeted and sustained operational performance of projects. Figure 1.10 identifies the regional infrastructure plans for the devolved regions.

Taking an over-arching view, the regional infrastructure schemes synergise with national infrastructure schemes. Notably the regional infrastructure schemes focus on city and growth deals across the major cities within each region. This means that infrastructure projects regionally have a local context and contribute to the Government's fair and levelling up infrastructure investment strategy.

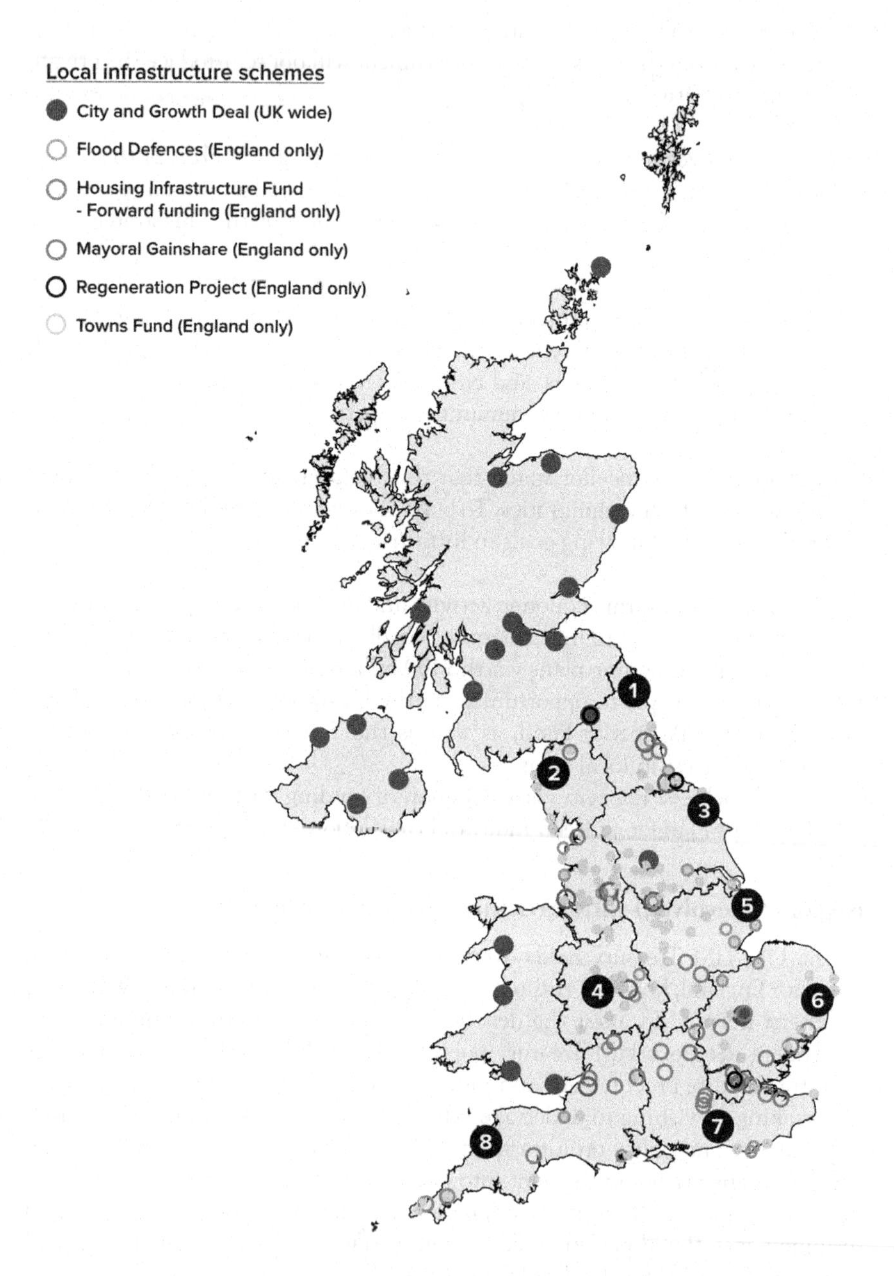

Figure1.10 Regional Infrastructure Plan. 1 North East; 2 North West; 3 Yorkshire and Humber; 4 West Midlands; 5 East Midlands; 6 East of England; 7 London & South East; 8 South West

Source: *Fairer, Faster, Greener* – HM Treasury (2020).

Summary

Global society is under great pressure to meet the infrastructure investment challenges (Bhattacharya et al., 2016 cited in OECD, 2017). Public investment in infrastructure will play a major role in the context of the post-pandemic world. Both economic and social infrastructure will be required to ensure resilience. However, there is still a lack of investment from governments and, currently, there is the lowest private sector investment in infrastructure in 10 years, which is partly to do with infrastructure not being visible as an asset market. Furthermore, sustained infrastructure investment could support recovery from the pandemic through the creation of a thriving island where people want to live, work and invest sustainably.

There is also an emerging friction between national, regional and local infrastructure investment. This book will explore these frictions and identify opportunities for investing nationally, locally and in collaboration with private investors.

Investing in infrastructure as a stimulus requires guidelines on investing in the infrastructure of tomorrow. However, there remains a need to plan, procure and deliver infrastructure better. The next chapter considers models of infrastructure investment funding, nationally, locally, and collaboratively, leading to, in Chapter 3, models for procuring infrastructure.

References

Adetola, A., Goulding, J. and Liyanage, C. (2011). "Collaborative engagement approaches for delivering sustainable infrastructure projects in the AEC sector: A review", *International Journal of Construction Supply Chain Management*, vol. 1, no. 1, pp. 1–24.

Goodwin, B. (2020, November). "Infrastructure strategy finally published – was it worth the wait?". 25 November. Available at: https://www.ice.org.uk/news-and-insight/the-infrastructure-blog/november-2020/national-infrastructure-strategy-published

Grimsey, D. and Lewis, M.K. (2002) "Evaluating the risks of public private partnerships for infrastructure projects", *International Journal of Project Management*, vol. 20, no. 2, pp. 107–118.

HM Treasury (2020, November). *National Infrastructure Strategy: Fairer, Faster, Greener*, CP 329 (London: HM Treasury). Available at: https://assets.publishing.service.gov.uk/government/uploads/system/uploads/attachment_data/file/938049/NIS_final_web_single_page.pdf

HM Treasury (2020, November). *UK Spending Review*, CP330 (London: HM Treasury). Available at: https://assets.publishing.service.gov.uk/government/uploads/system/uploads/attachment_data/file/938052/SR20_Web_Accessible.pdf

HM Treasury (2021, March). Budget 2021: *Protecting the Jobs and Livelihoods of the British People*, HC 1226 (London: HM Treasury). Available at: https://assets.publishing.service.gov.uk/government/uploads/system/uploads/attachment_data/file/966868/BUDGET_2021_-_web.pdf

HM Treasury (2021, March). *Build Back Better: Our Plan for Growth*, CP401 (London: HM Treasury). Available at: https://assets.publishing.service.gov.uk/government/uploads/system/uploads/attachment_data/file/968403/PfG_Final_Web_Accessible_Version.pdf

Howes, R. and Robinson, H. (2005). *Infrastructure for the Built Environment: Global Procurement Strategies*, 1st edition (Oxford: Butterworth-Heinemann).

Infrastructure and Projects Authority (2020).*Analysis of the National Infrastructure and Construction Procurement Pipeline 2020/21* (London: HM Stationery Office).Available at: https://assets.publishing.service.gov.uk/government/uploads/system/uploads/attachment_data/file/892451/CCS118_CCS0620674232-001_Pipeline_document_2020_WEB.pdf

McKinsey Global Institute (June, 2016). "Bridging global Infrastructure gaps", McKinsey & Company, in collaboration with McKinney's capital projects and infrastructure practice.

OECD (2015, February). *OECD Economic Surveys 2015: United Kingdom*. Available at: https://www.oecd.org/economy/surveys/UK-Overview-2015.pdf

OECD (2017). *Investing in Climate, Investing in Growth*. Paris: OECD Publications. Available at: https://www.nab.vu/sites/default/files/documents/9717061e.pdf

OECD (2020). *G20/OECD Report on the Collaboration with Institutional Investors and Asset Managers on Infrastructure: Investor Proposals and the Way Forward*. Available at: https://www.oecd.org/daf/fin/private-pensions/Collaboration-with-Institutional-Investors-and-Asset-Managers-on-Infrastructure.pdf

OBR (2020, March). *Economic and Fiscal Outlook – March 2020*, "Public sector net investment", Box 4.1, p. 162 (London: Office for Budget Responsibility). Available at: https://obr.uk/box/public-sector-net-investment/

OBR (2021, April). "Monthly public finances" (London: Office for Budget Responsibility, HM Treasury). Available at: https://obr.uk//docs/dlm_uploads/April-2021-PSF-commentary.pdf

ONS (2018, August). "Developing new statistics of infrastructure" (London: Office for National Statistics). Available at: https://www.ons.gov.uk/economy/economicoutputandproductivity/productivitymeasures/articles/developingnewmeasuresofinfrastructureinvestment/august2018

ONS (2021, December). *Public Sector Finances, UK: November 2021* (London: Office for National Statistics). Available at: https://www.ons.gov.uk/economy/governmentpublicsectorandtaxes/publicsectorfinance/bulletins/publicsectorfinances/november2021

Oxford Economics (2017). *Global Infrastructure Outlook – Infrastructure Investment Needs* (Oxford: Oxford Economics).

Rutherford, A. (2002). *Routledge Dictionary of Economics*, 2nd edition (London and New York: Routledge).

TUC (2021, May). *Ranking G7 Green Recovery Plans and Jobs*. Available at: https://www.tuc.org.uk/sites/default/files/2021-05/TUC%20G7%20Green%20Recovery%20Ranking%20report.pdf

World Economic Forum (2013). *Green Investment Report* (Switzerland: WEF). Available at: https://www3.weforum.org/docs/WEF_GreenInvestment_Report_2013.pdf

World Economic Forum (2019). *Global Competitiveness Report*, "Pillar 2" (Switzerland: WEF). Available at: https://www3.weforum.org/docs/WEF_TheGlobalCompetitivenessReport2019.pdf

World Bank (2012). *Best Practices in Public–Private Partnerships in Latin America: The Role of Innovative Approaches* (Washington, DC: World Bank).

World Bank (2020). World Bank Data. Available at https://data.worldbank.org/indicator/NE.GDI.FTOT.ZS

2 National investment in infrastructure

Introduction – infrastructure finance and funding

Delivering sustainable infrastructure requires substantial investments. The Paris Climate Change Accord, the UN Sustainable Development Goals, the Habitat III New Urban Agenda and the Sendai Framework for Disaster Risk Reduction have all shown the need for a more strategic approach to investing in public infrastructure, one that leverages private and institutional capital more effectively.

A range of finance models and funding opportunities across multiple sectors will be required. Furthermore, these financial and investment requirements will need to be developed and supported by governments through new and enhanced business models and finance structures. The multiple sectors' aspect of the new and enhanced finance and investment opportunities should seek to deliver inclusive growth by investing in national and regional sectors.

Learning outcomes

In this chapter you will learn about:

1 Key sources of infrastructure investment including trends in public and private financing of infrastructure projects.
2 The role of Government as key investor and sponsor of national infrastructure.
3 The growing need to expand private sector investment into national infrastructure programmes and the key private sector investors of infrastructure.
4 The emerging green financing options that help to deliver sustainable infrastructure.

Key sources of infrastructure Investment

Emanating from after World War II, the public sector has been responsible for the direct funding and long-term financing of infrastructure assets and services. It finances most infrastructure projects through both direct capital investment and private sector collaborations. And, although infrastructure is mostly financed traditionally using public funds, in financial terms it is considered a capital

DOI: 10.1201/9781003171805-2

investment/expenditure rather than a public sector expense. According to Standard & Poor's (2014), globally governments invest 3 per cent of GDP in infrastructure annually.

During the 1980s many countries, including the UK, embarked on the path of privatisation. To attract private investors required Government to set market conditions for attracting capital through effective regulation, budgeting, budget and investment forecasting, and project development and management. Private investors were encouraged to engage in the programme of privatisation to improve the infrastructure assets, which was seen to receive a 'reasonable' return on the capital they invested. The introduction of privatisation into infrastructure took on the collaborative name of public–private partnerships (PPP or P3). The use of PPP expanded greatly in the 1990s, with many more countries adopting this collaborative approach to finance.

Currently project finance is much more complicated than that publicly funded or funded through a PPP. There now exists a range of public and private finance options, as detailed in Figure 2.1.

Figure 2.1 identifies the key sources of finance for infrastructure– public finance and private finance. Public sector sources of finance include national, subnational and development finance such as infrastructure banks and the European Investment Bank. Private finance is considered corporate and project-related. Corporate finance is a blend of public and private companies, and refers to balance sheet expenditure, whereas project finance either follows a PPP model or a non-PPP model.

On-balance sheet Government debt instruments, such as bonds or direct project finance, will continue to be effective finance solutions where balance sheets are strong. Conversely off-balance sheet special purpose vehicles (SPVs), in collaboration with private entities, may be one solution for addressing balance sheet constraints. However, this will vary in relation to Government accounting practices. In many cases, assets developed by Government, or those acquired through

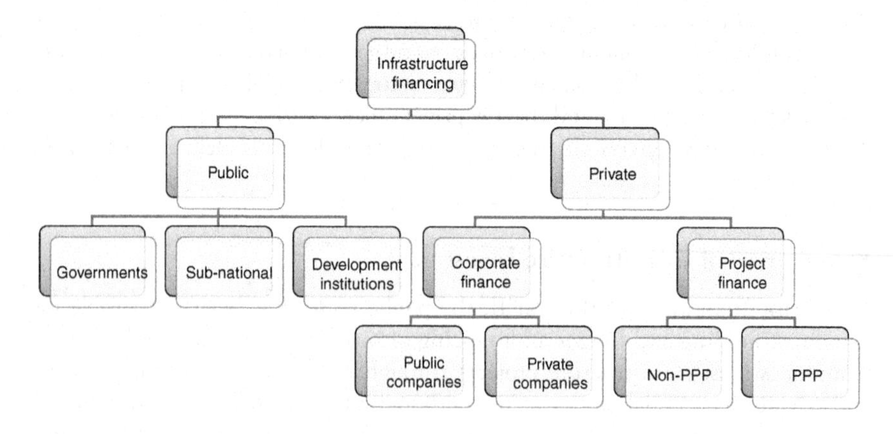

Figure 2.1 Infrastructure financing
Source: Inderst (2016).

private finance initiative contracts (including PPPs), should be kept on-balance sheet and the associated borrowing recorded as a liability.

It should be noted that the terms financing and funding are distinct. Financing is the provision of upfront capital sources from public, private, or collaborative suppliers of finance. Whereas the funding refers to who pays, i.e., the users/consumers (via fees and charges) or taxpayers or a combination of both.

National investment in infrastructure

Today, investment in infrastructure is still dominated by the public sector, of which traditional sources of funding are based on raising revenue streams through taxation. National taxation revenue streams in the UK include income taxes, corporation tax, sales tax (including value added tax) and property taxes which include business rates, council tax and stamp duty land tax. The cost of funding the long-term operation of the asset, combined with the increasing demand to invest more in infrastructure, places a significant burden on the public sector finances. Therefore, public budgetary constraints that result from a combination of factors, such as rising sovereign debt and slower economic growth, lead to insufficient public finance for capital investment.

The capital investment into infrastructure uses a variety of schemes including the UK Guarantees Scheme, national development banks and multinational development banks.

UK Guarantees Scheme

The UK Guarantees Scheme was introduced in 2012, with the first infrastructure project guaranteed in April 2013. The scheme was brought in to facilitate the delivery of infrastructure projects that were stalling due to the private sector investors facing difficulties in securing credit. The scheme was originally intended to expire in 2016 but has now been extended to 2026. The Treasury committed financial support to projects considered eligible, with eligibility determined by the definition of infrastructure set out in the Infrastructure (Financial Assistance) Act 2012, but in reality the HM Treasury consider eligibility to be 'nationally significant' projects.

The Treasury guarantees that lenders will be repaid, in full, for their investment in the infrastructure project. The scheme guarantees the principal and interest payments on infrastructure debt issued by the borrower to banks and investors. These guarantees are issued on a commercial basis. As illustrated in Figure 2.2, the private company pays a fee directly to the Treasury for its guarantee and, in return, the Treasury assumes the company credit risk. The fee paid by the company is determined by the Treasury's assessment of project risk and the price of this equivalent risk in the market at the time.

The provision of the scheme created opportunities, through the guarantee, for a wider pool of investors to invest in infrastructure. It has enabled a modest amount of guaranteed finance (around £1.8bn) for 9 infrastructure projects (see Figure 2.3) and 'prequalified support' for 39 further projects (around £34bn).

Note

1 The company agrees to reimburse HM Treasury for any payments under the UK Guarantees Scheme. HM Treasury and the company document various bilateral rights.

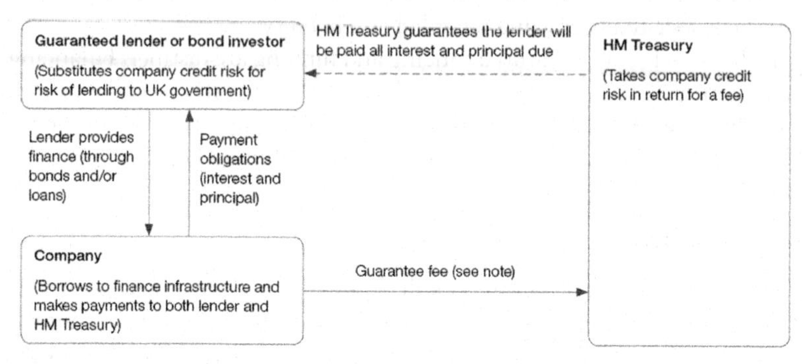

Figure 2.2 UK Guarantees Scheme
Source: National Audit Office (2015).

Project Name	**Sector**	**Scheme Status**
Drax Power	Energy	Signed (£75m)
Sustainable Development Capital - UK Energy Efficiency Investments Fund	Transport	Signed (£8.8m)
Northern Line Extension	Transport	Signed (£750m)
Mersey Gateway Bridge	Transport	Signed (£257m)
Ineos Grangemouth Ethane Import and Storage Facilities	Transport	Signed (€285m)
Speyside CHP Plant	Energy	Signed (£48.2m)
University of Northampton	Education	Signed (£292m)

Figure 2.3 UK Guarantees Scheme projects
Source: HM Treasury (2017).

The scheme means that the Treasury was in effect the second most active lender to new UK infrastructure projects in 2014. The National Audit Office (NAO), however, has been critical about the value for money that these guarantees offer the taxpayer. Specifically, the NAO raised concerns about the 'genuine need' of the projects and whether the delivery of them would be of significant benefit to society. Although there have not been any calls on the guarantees issued under the current scheme, the Government has in the past paid out on guarantees it has provided to infrastructure investments. These have included the High Speed 1 project and Metronet (London Underground public–private partnership). In June 2021 the administration and responsibility for the UK Guarantees Scheme moved to the UK Infrastructure Bank.

The Government also provides guarantees in other areas. For example, in 2020 the Government introduced a guarantee scheme for new-build affordable housing called the Affordable Homes Guarantee Scheme (AHGS). Here, loans will be funded by a capital markets bond programme which will have the benefit of a guarantee from the Ministry of Housing, Communities and Local Government (MHCLG).

National development banks

National development banks (NDBs) are Government-supported financial institutions. NDBs have existed since the nineteenth century, and by 2018 NDBs held 6.5 per cent of global banking assets (World Bank, 2021).

NDBs have specific policies pertaining to development funding, including long-term equity stakes and favourable credit agreements, and can borrow from other institutions or issued debt in domestic markets. As with the case of the UK Infrastructure Bank, NDBs provide long-term finance to early-stage private investors. What is different about the UK Infrastructure Bank (see below) from other NDBs is its infrastructure mandate. According to the Coalition for Urban Transition (2017) –a survey of 90 NDBs across 61 countries– only 4 per cent have an infrastructure mandate.

UK Infrastructure Bank (UKIB)

In November 2020 the Chancellor of the Exchequer, Rishi Sunak, announced in his autumn statement that the Government would establish a UK Infrastructure Bank (UKIB; HM Treasury, 2020). The function of the bank was documented in HM Treasury's design publication published in March 2021. The rationale for the establishment of the UKIB was to fulfil Government election promises of 'building back better, greener and fairer'. The HM Treasury (2021) publication identified the following two key objectives for a national infrastructure bank:

1. Help tackle climate change, particularly meeting our net zero emissions target by 2050.
2. Support regional and local economic growth through better connectedness, opportunities for new jobs and higher levels of productivity.

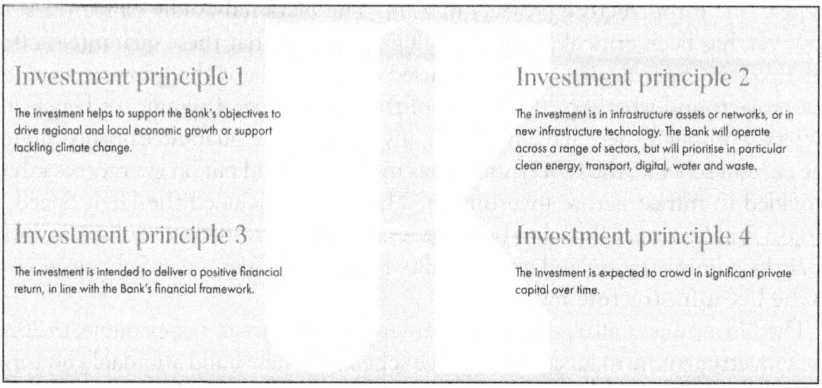

Investment principle 1

The investment helps to support the Bank's objectives to drive regional and local economic growth or support tackling climate change.

Investment principle 2

The investment is in infrastructure assets or networks, or in new infrastructure technology. The Bank will operate across a range of sectors, but will prioritise in particular clean energy, transport, digital, water and waste.

Investment principle 3

The investment is intended to deliver a positive financial return, in line with the Bank's financial framework.

Investment principle 4

The investment is expected to crowd in significant private capital over time.

Figure 2.4 UK Investment Bank –four investment principles
Source: UKIB (2021).

The UKIB (2021) offers private investors and local authorities 'loans, credit enhancement and equity investments for projects'. These capital investment offers will be made on a case-by-case project evaluation. To assist investors with the project evaluation, the UKIB (2021) provides the following four investment principles that projects and investors must satisfy, as shown in Figure 2.4.

Private investment is vital for the UK to address the infrastructure investment gap. However, attracting private investors is more than just sourcing and securing funding for infrastructure projects; private investment can also add 'expertise and capacity to governments and help them to realise their plans' (HM Treasury–*Policy Design*, 2021). The UKIB will provide a range of financial tools and products that will help deliver on its mandate (see Figure 2.5).

The establishment of the UKIB is a direct consequence of the Brexit deal, specifically the removal of the European Investment Bank (EIB) as a key funding institution. It is proposed that this new investment bank will fill the investment gap left by the EIB and move quickly on securing funding and long-term investment for national infrastructure projects, particularly the UK's offshore windfarms. By signalling investment in renewable energies, the UKIB aims to play a significant role in supporting the transition of the UK's economy to net-zero emissions by 2050. However, there is a lack of clarity in the UKIB framework document on the delivery of the net-zero carbon emissions target. Despite this pessimism, the overarching aim of the UKIB is to support the private sector market by 'taking a long-term approach to its inventions to provide the market with confidence and clear direction'.

Private investment is critical to the achievement of the UK's infrastructure ambitions. In supporting private sector investors, the *National Infrastructure Strategy*

Financing tools

4.2 The Bank will have a range of financing tools at its disposal so it can invest across the capital structure and will deploy them according to the needs of the particular project. The Bank will be able to offer the range of products below from inception.

4.3 **Senior debt:** this form of direct lending can be used as a tool to respond to low levels of liquidity in the market or overall capacity issues for larger projects, particularly in their construction phase.

4.4 **Hybrid products**: the Bank can provide mezzanine loan instruments, which could act as a first-loss product sitting between equity and senior debt, or other forms of credit enhancement. These can improve the efficiency of a project's capital structure and the related weighted average cost of capital.

4.5 **Equity:** equity can also be used to address construction risk or to assist in crowding in additional investors. The Bank will be able to make direct equity investments, as well as co-investing through funds.

4.6 **Guarantees:** the Bank will take on the UK Guarantees Scheme (UKGS) to provide guarantees where these are more efficient than the funded products above. The Bank will be able to use UKGS in a flexible way, including considering first-loss guarantees.

Figure 2.5 UK Investment Bank –financing tools
Source: HM Treasury–*Policy Design* (2021).

2020 advocates that in addition to the formation of UKIB, Government will also provide:

- Clear and enduring framework for investors and businesses
- Produce a policy paper on economic regulations and how to inject more competition into strategic investments
- Develop new revenue support models and remain open to new ideas in the market.

Multinational development banks

Multinational development banks (MDBs) are important sources for mobilising capital for infrastructure investments (especially in low-income countries). MDBs are charted by two or more countries (developed and developing) to improve economic recovery and output. In addition to direct debt and equity finance, MDBs can provide loan guarantees; offer in-house project preparation and technical project appraisal; undertake deal structuring; and generally support developers through high-risk phases. The targeted application of MDB guarantees to address risks that can be critical in the success of large infrastructure projects.

Annual infrastructure financing from MDBs more than doubled from 2004 to 2013, from US$20 billion to about US$54 billion (New Climate Economy, 2016) and more recently the Asian Development Bank (ADB) has increased its lending capacity by 50 per cent. MDBs are also committed to climate finance; for example,

> the ADB is committed to ensure that at least 75 per cent of its committed operations (on a 3-year rolling average) will be supporting climate change mitigation and/or adaptation by 2030 and climate finance from ADB's own resources to reach $80 billion cumulatively from 2019 to 2030. In 2021, ADB announced it was elevating its climate finance ambition to $100 billion by 2030.
>
> (ADB, 2022)

However, many see MDBs as operating below their potential. Leveraging some of their unique features could allow them to play a much larger role moving forward. Doing so may require more flexibility (e.g. in gearing ratios or willingness to accept a lower credit rating than AAA).

MDBs have been key sources for increasing capital investment over existing alternatives finance tools. Crucially they have contributed to underpinning some developing countries, in order to overcome a number of barriers and failures to access private financial markets.

Studart and Gallagher (2018) reported that Multilateral Development Banks (MDBs) and other authorities are currently targeting their investment into green infrastructure for the purposes of addressing the new wave of global climate change. MDBs, including the World Bank, the Asian Development Bank (ADB) and other developing financial institutions have increased their activities – annual infrastructure financing from MDBs more than doubled from 2004 to 2013, from US$20 billion to about US$54 billion.

European Investment Bank (EIB)

A central part of the EIB is the European Fund for Strategic Investments (EFSI). From 1973–2017 the UK contributed 16.11 per cent capital to the bank. And yet, following the UK's withdrawal from the EU, the UK no longer qualifies for this source of funding. However, funding committed prior to the UK's withdrawal from the EU is still available. Figure 2.6 highlights the volume of investment received by the UK from the EFSI.

As mentioned above, in direct response to the loss of this source of finance, the UK Government established the UK Infrastructure Bank (UKIB). The public sector unveiling of the UK Guarantees Scheme and the establishment of the UKIB are clear signals the public sector cannot, nor desires, to entirely fund and manage public infrastructure assets. More importantly the debate on how infrastructure should be financed has moved to who should fund infrastructure. The

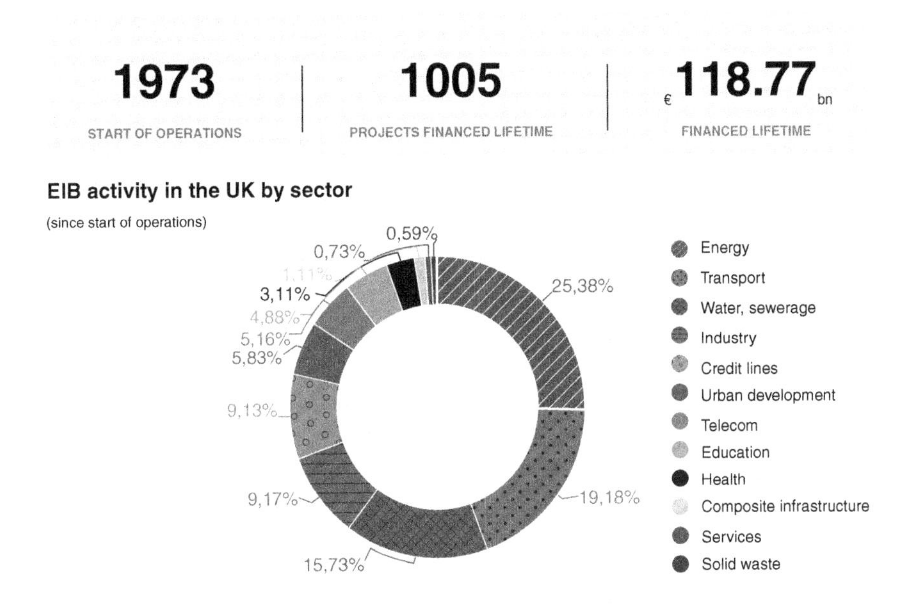

Figure 2.6 European Investment Bank, UK investments
Source: EIB (2021).

infrastructure investment gap has created an opportunity for investors and Government to gain access to the necessary additional finances required to deliver the investment in sustainable urban infrastructure.

The private investor

The recent global Covid-19 pandemic and resultant historic levels of government debt, have forced many governments to reconsider how best to finance infrastructure projects, and in doing so bolster the economy, improve the standard of the infrastructure asset and improve the quality of life for all. Rethinking infrastructure investment has given the private sector an opportunity to provide both short-term and long-term financing for infrastructure projects.

Since the global economic recession in 2008, many private institutional investors have reassessed their portfolios and been drawn to infrastructure projects as a new 'asset class' for several reasons, as described by Inderst (2010) below:

1 The long-term nature of the investment;
2 Stability of the cash flows with public sector actors' involvement as clients or guarantors;
3 Less irregular returns; and the
4 Provision of an essential service (digital networks, utilities, healthcare or education) with a natural monopoly within a geographical area.

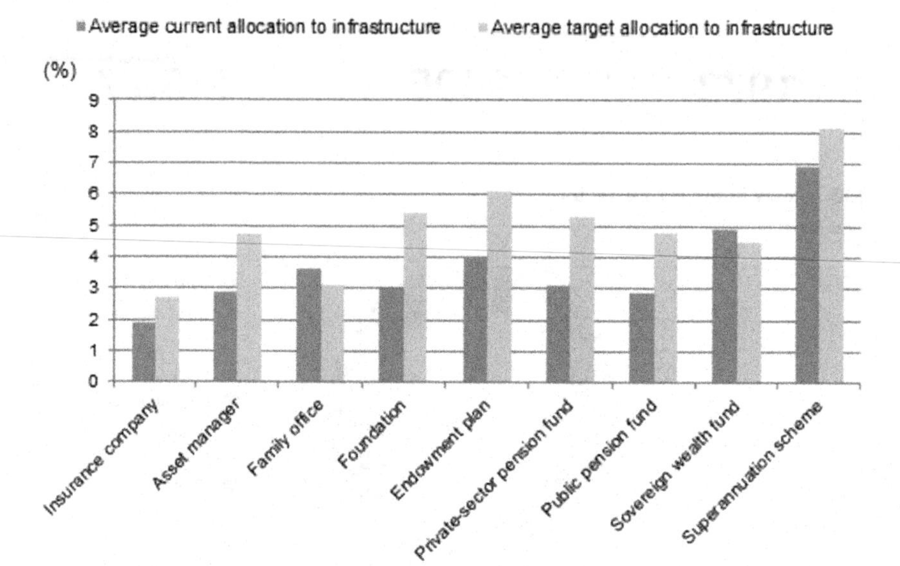

Figure 2.7 Breakdown of infrastructure by investor type
Source: Standard & Poor's (2014).

Globally, private investment in infrastructure has remained relatively stable and resilient to the Covid-19 pandemic. According to GI Hub (2021b), 'in 2020 private investment in infrastructure was valued at USD156 billion, representing 0.2% of total global GDP and far shy of the 5% of GDP (combining public and private investment) that some studies show is required to close the infrastructure gap'. Yet, private investors could reduce the infrastructure deficit gap and significantly contribute to the UK's desire to 'build back better' and, in response to this growing dependency, a range of financial instruments and mechanisms are now available to private investors, with different instruments being attractive to diverse types of investors (see Figure 2.7).

Hence, the private investors' role is clear – to provide short and long-term finance for infrastructure investment. At the start of the current century, Forrester (2001) found that, together with the conventional role of commercial banks in project finance transactions, new services were emerging for the infrastructure projects industry, namely:

1 Construction financing,
2 Working capital finance for infrastructure projects globally,
3 Financial consultancy services, and
4 Acting as financial intermediators for-long term fixed-rate financing.

Whereas short-term finance is supplied for businesses' day-to-day operations, long-term finance of infrastructure assets is equally important insofar as it secures the long-term viability of the project which in turn enhances the economy through employment, growth and competitiveness. Additionally long-term finance transfers the risks associated with the investment to the funding supplier (i.e. global institutional investors, capital markets and banks). This is due to the inherent fluctuations in financial markets, such as interest rate fluctuations. The transfer of risk in relation to short-term finance is transferred to end-user creditors (i.e. governments and corporations) as the result of their obligation to continuously liquidate their debt (Peria and Schmukler, 2017).

Private capital can be channelled into infrastructure through two forms: corporate financing and project financing. Project financing is characterised generally as non-or limited-recourse lending for the development of large, highly specific, capital-intensive projects. By lending directly to the project by means of a special purpose vehicle (SPV), the assets of the SPV are leveraged as collateral for the loan with the liability being limited to the invested capital.

Project financing capital can be debt or equity. Nowadays private investors are considered a key source of infrastructure finance with the dominant, private sector investors being commercial and investment banks, resulting in 80 per cent of private investment in infrastructure projects as debt finance. Although project lending is capital intensive, debt financing is an attractive instrument for many reasons in that it is usually tax deductible and it can be kept off the balance sheets (Tan, 2007). Typically characterised as fixed-income, project debt financing can be filtered into either greenfield or brownfield investments:

- Greenfield project debt financing is normally supplied in two forms: project loans or project bonds. Conventionally, most project debt is supplied in the form of project loans provided by commercial banks who are willing to absorb construction risk in exchange for a higher rate of return (Gardner and Wright, 2011). 'A single bank may supply all capital requirements or, alternatively, as a syndicate by a Mandated Loan Arranger (MLA) to comprise layered or mezzanine financing which typically offers higher interest rates and increased equity rates' (Tan, 2007).
- Brownfield project debt is usually post construction completion where other entities such as institutional investors, real estate investment trusts (REIT) and sovereign wealth funds can invest into long-term bonds issued by the SPV after the asset has been commissioned. With lower risk tolerance than commercial banks, these nonbanking institutions prefer to enter post construction completion where the risk profile has been reduced and typically seek stable, medium to long term assets to diversify equity holdings (Della Croce et al., 2015).

Project financing can equally be suppled in the form of equity. Typically, project 'sponsors' are responsible for the capitalisation of project financing in the form of equity. Equity generally accounts for around 10 to 30 per cent of the finance though this can change in dynamic economic environments. Third-party investors are now actively participating in infrastructure investment by providing a portion of the equity capital. These have tended to be insurance companies, pension funds, wealthy individuals and trade union funds who have a portfolio of assets and participate with the intention of diversifying risks (Tan, 2007).

Corporate financing constitutes a private operator funding a project on-balance sheet. For example, private corporations and institutions provide private equity and infrastructure funds and seek the highest return and hence invest mostly in unlisted equity of projects with strong growth potential. Therefore, these funds often invest in relatively new or unproven markets and technologies. In contrast, pension funds and insurance companies search for investments that provide steady, long-term predictable income streams to meet long-term commitments.

Equity infrastructure investment

The term equity refers to capital supplied by institutional investors in exchange for dividends (i.e., share payments) or ownership of the asset. Equity financing therefore refers to companies selling stocks or infrastructure assets to investors to raise capital. Private equity (PE) funds are pools of capital that are used to invest into companies. The investment can be a direct investment into the company through a private equity fund/manager.

Most of the PE funds are structured as Limited Partnerships (LPs). PE funds are managed by the General Partners (GPs), who are usually large-scale institutional investors with potentially unlimited responsibilities towards such funds (Gilligan and Wright, 2014). This results in GPs receiving annual management fees, typically around 1.5–2 per cent of the committed capital. These investors provide most of the capital along with LPs, typically over a period of ten years (Døskeland and Strömberg, 2018). Figure 2.8 shows the PE investment cycle.

Since equity investors prefer to enter the project after construction completion in the presence of a reduced risk profile, these funds have tended to concentrate their investments on brownfield projects in mature markets such as Europe and

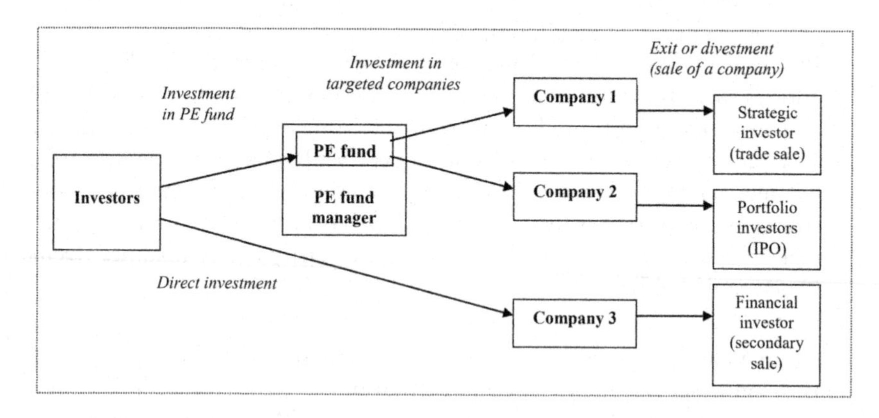

Figure 2.8 Private equity investment cycle
Source: Herasym and Segura (2006).

North America (Della Croce and Gatti, 2014). Equity infrastructure investment enables these third-party funds to diversify their investment portfolios. Moreover, the increased returns from equity outlay have further supported this form of investment as an attractive proposition. Resultantly, over the previous two decades, increased attention for infrastructure has led to the separation of 'infrastructure assets' from the 'alternative investments' category and, instead, it is now considered an asset class with its own distinct investment characteristics (Bielenberg et al., 2016).

Equity investors can access infrastructure either through listed or unlisted instruments. Insurance companies, pension funds and other private equity funds are now actively investing in unlisted infrastructure equity. The long-term investment horizon of these investors suggests they are a natural match to the infrastructure asset class.

Advantageously, private finance can negate budgetary pressures on public capital and provide a means to overcome funding shortfalls on the part of the public sector. Furthermore, it can act as a conduit for greater investment diversification of national investment portfolios as it enables capital flow into sectors which may not be available elsewhere (Pisu et al., 2015).

Case Study – Thames Tideway Tunnel

The aim of the Thames Tideway Tunnel infrastructure project is to alleviate the sewage overspill (approximately 39 million tonnes) into the Thames. The project consists of the construction of a 25km long interceptor sewage tunnel being constructed in central London.

The project at a glance:

Cost: £4.2bn

Dates: Construction commenced in 2016, works to complete in 2025.

Key goals:

- Upgrades to five sewage treatment works at a cost of £675m, largely completed in 2014;
- The 6.9km long Lee Tunnel at a cost of £635m, completed in 2015; and
- The 25km Thames Tideway Tunnel at a cost of £4.2bn.

Project funding:

The project is privately financed by the Bazalgette Consortium (made up of Allianz, Amber Infrastructure, Dalmore Capital and Dutch Infrastructure

Fund). The project is also supported by the UK Government's fiscal support package to mitigate the project risks and make the project viable for private financing.

The Water Services Regulatory Authority (Ofwat) was appointed the independent economic regulator to determine the charges to be borne by the customers of Thames Water for funding the tunnel construction.

Project benefits:

According to Tideway (2017–), the project will bring a 'rejuvenated river economy and a new public realm for the people of London' through investment in local communities, education, training and the supply chains.

Case Study – Belgrade Airport

Belgrade Airport is Serbia's main international airport, carrying most of the country's air traffic. The Belgrade Airport project was awarded to VINCI Airports SAS for a period of 25 years to develop, operate and manage Belgrade Airport. VINCI Airports, part of the VINCI S.A. group, is the world's fourth-largest airport operator.

The project at a glance:

Cost: EUR 980 million

Dates: Construction commenced in 2020, works are due to be completed in 2022.

Key goals:

- Expansion and reconstruction of the terminal building
  - 1, 200 m^2 of space for new restaurants, bars and shops
- Expansion of aprons
- Expansion works including construction on an area of nearly 25, 000 m^2
  - extension of passenger terminal – Pier C
  - 8 additional boarding gates equipped with air bridges
  - 5 gates for boarding passengers at open parking positions
  - additional roof corridor for arriving passengers
- New de-/anti-icing pad
- Inserted (new) runway
- Tesla parking
  - cargo terminal on an area of 36.400 m^2
  - More than 1, 500 parking spaces; 798 for the passengers (including parking spaces for people with reduced mobility), 109 for taxi services, 330 for employees and 265 for rent-a-car vehicles.
- Modernisation of access roads and car parks

Project funding:

The project demonstrates the benefit of private sector financing of infrastructure. The capex plan includes green investments, which ensure safer, more environmentally friendly and lower-cost of operation of the airport infrastructure throughout the entire concession. VINCI Airports officially took over the management of Belgrade Airport in December 2018 and signed a 25-year concession contract with Serbia covering the airport's financing, operation, maintenance, expansion, and modernisation. The Government of Serbia and Nikola Tesla Airport AD acted as the grantors of the concession. Project finance was provided by commercial banks and co-financed with other International Financial Institutions (IFC, Proparco, DEG.)

Project benefits:

Fundamentally the investment in Belgrade Airport expansion will deliver wide reaching economic benefits through increased international connectivity capacity. In turn, the increase in passenger numbers will also create and sustain jobs across the country.

Belgrade Airport (2020)

Pensions Infrastructure Platform (PiP)

The Pensions Infrastructure Platform (PiP) started out in 2012 as a UK Government initiative designed to support pension funds through investment in infrastructure. To date, PiP investment has been successful insofar as it has exceeded its initial funding objective by securing almost £750m of investment in infrastructure projects, predominantly social and green infrastructure. PiP has two investment funds, namely:

A multi strategy infrastructure fund – open to all UK pension schemes and designed to invest capital in infrastructure projects.
A single investment fund – providing a bespoke investment management service for pension scheme investors.

In August 2020 PiP was acquired by Foresight Group, an infrastructure and private equity manager investor, for £1.8bn. This merger has further enhanced the reputation of PiP as a UK infrastructure pension fund investment authority (Pensions Infrastructure Platform, n.d.).

Green investment

There is a great paradox to infrastructure investment; on one hand investment in infrastructure is an investment in the quality of life for citizens, yet according to the Global Infrastructure Hub, infrastructure consumes half of the world's

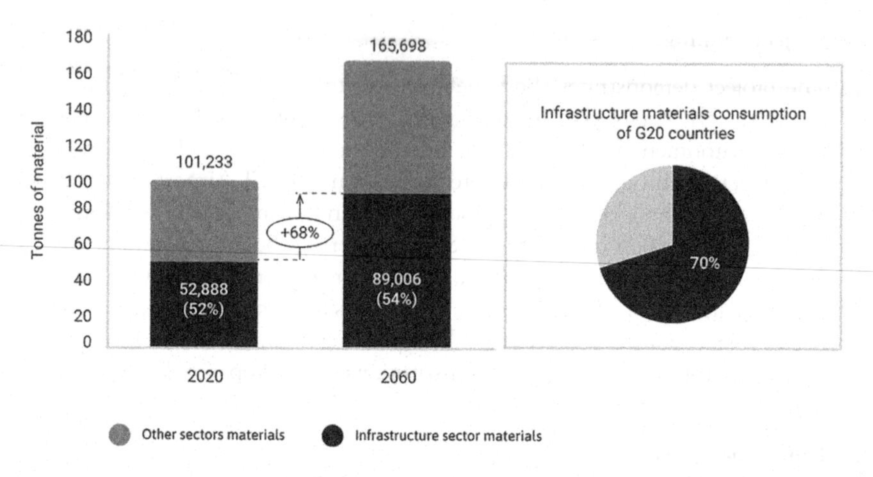

Figure 2.9 Global material consumption: infrastructure sector dominates consumption
Source: Global Infrastructure Hub (2021a).

projected materials annually (see Figure 2.9). This paradox can no longer be ignored by countries that have pledged to be carbon zero by 2050. So, whilst governments across the world are using the pandemic and the environmental crisis to justify a new splurge in infrastructure spending, there needs to be a greening up of these strategies (Monbiot, 2021), where a significant reduction in CO2 via alternative infrastructure practices could play a key role. This will also require dedicated green financial models and mechanisms.

Green Climate Fund

The Green Climate Fund (GCF), established in 2010, is a financial mechanism of the UN Framework Convention on Climate Change. The GCF works through a wide range of accredited entities that carry out a range of activities, from the development of funding proposals to the management and monitoring of projects and programmes. Financial inputs to the GCF Trust Fund are received in the form of grants, loans or capital contributions(GCF, 2020).

The international carbon finance market is driven by the Clean Development Mechanism (CDM) and Joint Implementation (JI), which were created through the Kyoto Protocol for 'converting' emissions reduction projects into Certified Emissions Reduction (CER) units, which can then be traded. The market mechanism is designed to channel project-based finance and carbon emission reduction technology from wealthier countries to lower-income countries. CERs can be generated through individual projects (e.g. a wind farm) or through an aggregation of interventions (e.g. installation of solar hot water heaters across a housing estate). Take-up has stayed low, however. Lack of awareness of existing registered programmes of activities that may be available for other parties to participate in, combined with transaction costs and complexity, remain substantial barriers.

Overarching goal: Every financial decision takes climate change into account

Reporting 	• Private sector to refine TCFD climate-related financial disclosure to increase quantity and quality of reporting	• Agree potential paths to mandatory reporting at domestic and international levels	• Build coalitions of countries who mandate reporting and companies committed to full climate disclosures
Risk Management 	• Assess the resilience of firm's strategies to net zero transition through stress tests	• Develop open source, business-relevant reference scenarios for regulators, financial firms and businesses to test strategic resilience	• Establish coalition of central banks and regulators committed to issuing guidance on risk management and running stress tests
Return 	• Enable investors to make informed decisions on whether companies and portfolios are transition ready	• Agree metrics to measure net zero/ alignment of investment portfolios	• Build coalition of financial institutions that commit to net zero alignment and measurement and disclosure of progress
MDBs/DFIs 	• Encourage MDBs to report their own emissions and exposure to climate risks, in line with TCFD	• Realise MDBs' commitments to transition plans to achieve Paris Alignment	• Explore rapid expansion of blended financing for climate resilience, adaptation and mitigation
Innovative Finance 	• Work with the private sector to promote the most promising and impactful financial innovations in sustainable finance (including transition bonds, contingent climate securitisations, and the scaling up of rapid private markets for carbon offsets and nature-based solutions)		

Figure 2.10 COP26 private finance strategy to drive whole economy transition
Source: Bank of England (2021).

G-7 leaders recognise that a new long-term strategy for resilient infrastructure is needed to close the infrastructure gap. A range of investment in public infrastructure is needed to kickstart the global recovery from the Covid-19 pandemic, to meet UN sustainability goals, and to mitigate the climate emergency in accordance with the 2015 Paris Agreement and protect biodiversity. This strategy includes strengthening regional partnerships to support sustainable finance and leveraging public finance to crowd-in private finance and this is reflected in the UK's *Build Back Better* strategy (HM Treasury, 2021; see Figure 2.10).

Green Investment Group (GIG)

Originally, the GIG was launched by the UK Government in 2012 as the Green Investment Bank (GIB). The GIB initiative aimed at securing private finance to fund green infrastructure projects and committed £2.8bn directly to such projects. The GIB also created an offshore wind fund, which invested in five operational wind farms, making it the largest renewable fund in the UK. On 18 August 2017, the sale of the GIB to the Macquarie Group was completed and it was renamed the GIG. According to the GIG progress report (2018), GIG targets investments in green infrastructure projects across:

- established technologies, including offshore and onshore wind, solar, hydro, inter-connectors, and waste and bioenergy;
- emerging technologies, including tidal, biofuels, energy efficiency, storage, low-carbon transport, smart grids, and district heating;
- all stages of the project lifecycle (development, construction, and operations); and
- the entire capital structure from equity to debt.

Case Study – Smart Energy Networks

In the city of Nice, France, a new district called "Nice Méridia" is providing over 500, 000m^2 of space, which includes housing, office space, retail, leisure, school, and a hospital. Nice determined that this new district should become a smart energy network.

The project at a glance:

Cost: Undisclosed

Dates: Commenced in 2018 with a 25-year concession contract.

Key goals:
Heating and cooling network (each of them 5.6km long).

A geothermal powerplant (up to 10MW Pmax, producing 80 per cent renewable energy), and smart grid supervision utilising:

- Geothermal energy
- Photovoltaic panels
- Energy storage for heat, cold and electricity

Project funding:

In 2018, Idex was awarded a contract for a period of 25 years to design, finance, realise, operate and maintain the production and distribution of heat and cold as well as to ensure energy efficiency by implementing a smart grid. Idex created a company dedicated to this project called Méridia Smart Energie.

Project benefits:

- 80 per cent of energy used will be renewable
- Reduction of greenhouse gas emissions by 20 per cent
- Reduction of energy bills by 40 per cent
- Create 60 full-time jobs during the construction phase and 5 full-time jobs during the operations and maintenance phase
- Increase the value of buildings

Global Infrastructure Hub (2021c–*Smart Energy Networks*)

Summary

Infrastructure investment can be sourced from either public or private options. The government may generate capital through taxes, or borrow directly from national and international institutions. The state may also leverage infrastructure financing from the private-sector. Predominantly, the Government has provided most of the funding for infrastructure provision either through taxes or borrowing warranted by the perceived socioeconomic benefits. However, in the current economic context, and the associated governmental capital retrenchment, there is appetite for greater mobilisation of private finance to alleviate the infrastructure financing gap.

Utilising private capital, infrastructure financing can serve to alleviate budgetary constraints by freeing up public sector resources that can be used elsewhere. Moreover, the long-term asset life of infrastructure places it as an attractive proposition for institutional investors and other alternative sources of capital.

References

Asian Development Bank (2022). "Climate finance in 2021". Available at: https://www.adb.org/news/infographics/climate-finance-2021

Bank of England (2021). "COP26 private finance strategy to drive whole economy transition", Bank of England. Available at: https://www.bankofengland.co.uk/-/media/boe/files/events/2020/february/cop26-private-finance-strategy.pdf

Belgrade Airport (2020). "The Project". Available at: https://beg.aero/eng/construction/the_project

Bielenberg, A., Kerlin, M., Oppenheim, J. and Roberts, M. (2016). "Financing change: How to mobilize private-sector financing for sustainable infrastructure". McKinsey Center for Business and Environment. Available at: https://newclimateeconomy.report/workingpapers/workingpaper/financing-change-how-to-mobilize-private-sector-financing-for-sustainable-infrastructure/

Coalition for Urban Transitions (2017). "Global review of finance for sustainable urban infrastructure" (Washington, DC: Coalition for Urban Transitions). Available at: https://newclimateeconomy.report/workingpapers/wp-content/uploads/sites/5/2018/01/NCE2017_CUT_GlobalReview_02012018.pdf

Della Croce, R. and Gatti. S. (2014). *Financing Infrastructure – International Trends* (Paris: OECD).

Della Croce, R., Paula, J. and Laboul, A. (2015). *Infrastructure Financing Instruments and Incentives* (Paris: OECD). Available at: https://www.oecd.org/finance/private-pensions/Infrastructure-Financing-Instruments-and-Incentives.pdf

Døskeland, T.M. and Strömberg, P. (2018). "Evaluating investments in unlisted equity for the Norwegian Government Pension Fund Global (GPFG)". Available at:https://www.hhs.se/contentassets/662e98040ed14d6c93b1119e5a9796a4/doskelandstromb-erg2018.pdf

Earth Negotiations Bulletin (2012). *Summary Report, 13–22 June 2012*. UNCSD (Rio+20). Available at: https://enb.iisd.org/events/uncsd-rio20/summary-report-13-22-june-2012

European Investment Bank (2021). "EIB activity in UK by sector", EIB, 2021. Available at: https://www.eib.org/en/projects/regions/united-kingdom/index.htm

Forrester, J.P. (2001). "Role of commercial banks in project finance". Available at: https://www.mondaq.com/unitedstates/project-financeppp-pfi/11272/role-of-commercial-banks-in-project-finance

Gardner, D. and Wright, J. (2011). "Project finance", HSBC. Available at: https://pdf4pro.com/view/david-gardner-and-james-wright-hsbc-126ac1.html

Gilligan, J. and Wright, M. (2014). *Private Equity Demystified: An Explanatory Guide*, 3rd edition (London: Business with Confidence, ICAEW Corporate Finance Faculty).

Herasym, H. and Segura, E.L. (2006, April). "The role and benefits of private equity in emerging market economies" (Houston, TX: The Bleyzer foundation, SigmaBleyzer).

Global Infrastructure Hub (2021a). "Global material consumption: Infrastructure sector dominates consumption". Available at: https://www.gihub.org/infrastructure-monitor/insights/infrastructure-consumes-more-than-half-of-the-world-s-materials/

Global Infrastructure Hub (2021b). *Infrastructure Monitor 2021: Private Investment in Infrastructure*. Available at: https://cdn.gihub.org/umbraco/media/4302/gihub_infrastructuremonitor2021_private_investment_in_infrastructure.pdf

Global Infrastructure Hub (2021c, 16 February). *Smart Energy Networks*. https://www.gihub.org/infrastructure-technology-use-cases/case-studies/smart-energy-networks/

HM Treasury (2021, March). *Build Back Better: Our Plan for Growth*. Available at: https://assets.publishing.service.gov.uk/government/uploads/system/uploads/attachment_data/file/968403/PfG_Final_Web_Accessible_Version.pdf

Green Climate Fund (2020–). Climate Funds Update. https://climatefundsupdate.org/.

Green Investment Group (2018, October). *Progress Report: A Year of Growth, Expansion and Becoming a Project Developer*. Available at: https://www.greeninvestmentgroup.com/assets/gig/corporate-governance/green-investment-group-progress-report-october-2018.pdf

Green Investment Group (n.d.). Website. https://www.greeninvestmentgroup.com/en.html

HM Treasury (2017). "UK Guarantees scheme: Table of prequalified projects" (London: HM Treasury). Available at: https://www.gov.uk/government/publications/uk-guarantees-scheme-prequalified-projects/uk-guarantees-scheme-table-of-prequalified-projects

HM Treasury (2020). *National Infrastructure Strategy* (London: HM Treasury). Available at: https://assets.publishing.service.gov.uk/government/uploads/system/uploads/attachment_data/file/938049/NIS_final_web_single_page.pdf

HM Treasury (2020, November). *Spending Review*. CP 330. Available at: https://assets.publishing.service.gov.uk/government/uploads/system/uploads/attachment_data/file/938052/SR20_Web_Accessible.pdf

HM Treasury (2021). *UK Infrastructure Bank Framework Document*(London: HM Treasury). Available at: https://assets.publishing.service.gov.uk/government/uploads/system/uploads/attachment_data/file/994437/UK_Infrastructure_Bank_Framework_Document.pdf

HM Treasury (2021). *UK Infrastructure Bank Policy Design* (London: HM Treasury). Available at: https://assets.publishing.service.gov.uk/government/uploads/system/uploads/attachment_data/file/994535/UKIB_Policy_Design.pdf

Inderst, G. (2010). Infrastructure as an asset class", *EIB Papers*, vol. 15, no. 1, pp. 70–104.

Inderst, G. (2016). "Infrastructure investment, private finance, and institutional investors: Asia from a global perspective". ADBI Working Paper 555, Asian Development Bank, Japan.

Infrastructure and Projects Authority (2016). *National Infrastructure Delivery Plan 2016–2021*. Available at: https://assets.publishing.service.gov.uk/government/uploads/system/uploads/attachment_data/file/520086/2904569_nidp_deliveryplan.pdf

Institute for Development and Transportation Policy (2015, April 10). "Two years in – how are the world's multilateral development banks doing in delivering on their $175 billion pledge for more sustainable transport?". Available at:https://www.itdp.org/2015/04/10/two-years-in-how-are-the-worlds-multilateral-development-banks-doing-in-delivering-on-their-175-billion-pledge-for-more-sustainable-transport/.

McKinsey Global Institute (2021, June 16), "Making infrastructure tech a reality in your portfolio'. Available at: https://www.mckinsey.com/business-functions/operations/our-insights/making-infrastructure-tech-a-reality-in-your-portfolio.

Monbiot, G. (2021). "We can't build our way out of an environmental crisis", *The Guardian*, 1 September.

National Audit Office (2015). *UK Guarantees Scheme for Infrastructure* (London: HM Treasury). Available at:https://www.nao.org.uk/wp-content/uploads/2015/01/UK-Guarantees-scheme-for-infrastructure.pdf. Also at: https://assets.publishing.service.gov.uk/government/uploads/system/uploads/attachment_data/file/209806/UK_Guarantee_-_A_brief_overview_-_Allen___Overy.pdfandhttps://www.gov.uk/government/publications/uk-guarantees-scheme-key-documents

New Climate Economy (2016, October). "Transforming the financial system". Section 3 of *The Sustainable Infrastructure Imperative Financing for Better Growth and Development* report. Available at: https://newclimateeconomy.report/2016/transforming-the-financial-system/

Pensions Infrastructure Platform (PiP) (n.d.). Available at:https://pipfunds.co.uk/ (accessed April 2022).

Peria, M.S.M. and Schmukler, S. (2017). "Understanding the use of long-term finance in developing economies", IMF WP17/96. Available at: https://www.imf.org/en/Publications/WP/Issues/2017/04/26/Understanding-the-Use-of-Long-term-Finance-in-Developing-Economies-44855

Pisu, M., Pels, B. and Bottini, N. (2015). *Improving Infrastructure in the UK* (Paris: OECD). Available at: https://www.oecd-ilibrary.org/docserver/5jrxqbqc7m0p-en.pdf?expires=1652451734&id=id&accname=guest&checksum=D7BE670AA424FBB2E2D2D0425F4D6204.

Standard & Poor's (2014, July). "Ratings Direct: Investing in infrastructure: Are insurers ready to fill the funding gap?" (New York: Standard & Poor's).

Studart, R. and Gallagher, K. (2018). 'Guaranteeing sustainable infrastructure', *International Economics*, vol. 155, pp. 84–91.

Tan, W. (2007). *Principles of Project and Infrastructure Finance*(London and New York: Taylor & Francis).

Tideway (2017–). *Reconnecting London with the River Thames: Delivering a Lasting Legacy* (London: Tideway). Available at: https://www.tideway.london/media/1624/tideway-legacy-brochure_2017.pdf

UK Infrastructure Bank (2021). "What we offer". Available at: https://www.ukib.org.uk/what-we-offer

World Bank (2018). "Increasingly popular national development banks must become more effective to solve challenges", World Bank Blogs. Available at: https://blogs.worldbank.org/psd/increasingly-popular-national-development-banks-must-become-more-effective-solve-challenges

3 Funding regional infrastructure

Introduction

This chapter examines the range of financial models and instruments used by local authorities across the UK to invest in regional infrastructure. Regional infrastructure investment is essentially and historically a Government-funded activity but more recently has included private investments directly (commercial lending) and indirectly (regional growth and city deals).

The structure of this chapter:

1 Local government investment
2 Devolved Government investment
3 City and regional growth deals

Learning outcomes

To understand the responsibilities of local Government in the delivery of regional infrastructure and investment.

Local government

In the UK local government is responsible for a range of essential services for citizens and businesses in defined geographical locations. These essential services include social care provision, schools, social housing and waste collection. According to the Local Government Information Unit (LGIU, 2021), there are five possible types of local authority in England (Figure 3.1). These are:

1 County councils – cover the whole county and provide 80 per cent of services in these areas, including children's services and adult social care, education, and highways and transport planning.
2 District councils – cover a smaller area within a county, providing more local services (such as housing, local planning, waste and leisure but not children's services or adult social care); can be called district, borough or city council.

DOI: 10.1201/9781003171805-3

3 Unitary authorities – just one level of local government responsible for all local services; can be called a council (e.g. Medway Council), a city council (e.g. Nottingham City Council) or a borough council (e.g. Reading Borough Council).
4 London boroughs – each of the 32 boroughs is a unitary authority.
5 Metropolitan districts – effectively unitary authorities, the name being a relic from past organisational arrangements. They can be called metropolitan boroughs or city councils.

In Scotland there are 32 local councils, in Northern Ireland there are 11 local councils, and there are 22 local authorities in Wales. These local regional councils are run by elected councillors. All these councils are unitary and operate a single-tier local government system.

Services provided by local government

Local Councillors are elected to represent people in a defined geographical area for a 4-year period. Councillors are required to determine local government policy and spend to deliver the needs of residents and businesses. Figure 3.2. identifies the range of services to be provided by local governments.

Funding of local government

Local authorities are funded from a variety of sources, including Government grants, council tax and fees and charges. Together, council tax and business rates

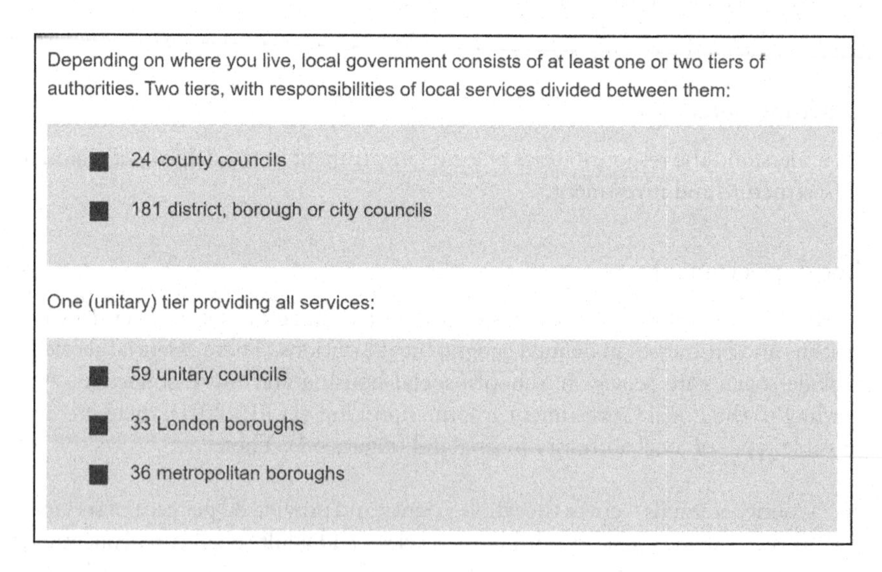

Figure 3.1 Types of local government in England
Source: Local Government Association (2021).

	Unitaries	County Councils	District Councils	Metropolitan Districts	London Boroughs	GLA
Education	✓	✓		✓	✓	
Highways	✓	✓		✓	✓	✓
Transport Planning	✓	✓		✓	✓	✓
Passenger Transport	✓	✓				✓
Social Care	✓	✓		✓	✓	
Housing	✓		✓	✓	✓	
Libraries	✓	✓		✓	✓	
Leisure and Recreation	✓		✓	✓	✓	
Environmental Health	✓		✓	✓	✓	
Waste Collection	✓		✓	✓	✓	
Waste disposal	✓	✓		✓	✓	
Planning applications	✓		✓	✓	✓	
Strategic Planning	✓	✓		✓	✓	✓
Local tax collection	✓		✓	✓	✓	

Figure 3.2 Local government service provision
Source: Local Government Information Unit (2021).

make up local authorities' largest source of income. Combined, local authorities account for almost one-quarter of all public sector spend (LGIU, 2021). Local authority spending can be divided into revenue expenditure and capital expenditure. Revenue expenditure (meeting employee costs) is financed through a balance of central Government grant, retained non-domestic (business) rates and the locally raised council tax. Capital expenditure (constructing or improving physical assets) is principally financed through central Government grants, borrowing and capital receipts.

In England, local government is funded through:

- grants from central Government (about 52 per cent) made up mainly of redistributed business rates, including the Revenue Support Grant and the Public Health Grant;
- and locally raised funding (about 48 per cent) which includes council tax (charged to local people), retained local business rates income, and other sources such as car parks, parking permits, rents and the hire of sports facilities.

Grant funding has decreased since 2015/16 (see Figure 3.3) and this has resulted in a decrease of spending per person in real terms as the result of this reduction and local councils being unable to fund the deficit. The UK Government has further committed to phasing out central grants for local government so that it will be funded entirely through local business rates and taxes.

As can be seen in Figure 3.4, 2019–20 the capital expenditure of England's local authorities was approximately £95 billion; the largest service spend was education services (£32.6 billion) and the lowest spend was on planning and development services (£1.4 billion). Across England as a whole, total spending on services fell by 4.3 per cent in real terms between 2015–16 and 2019–20.

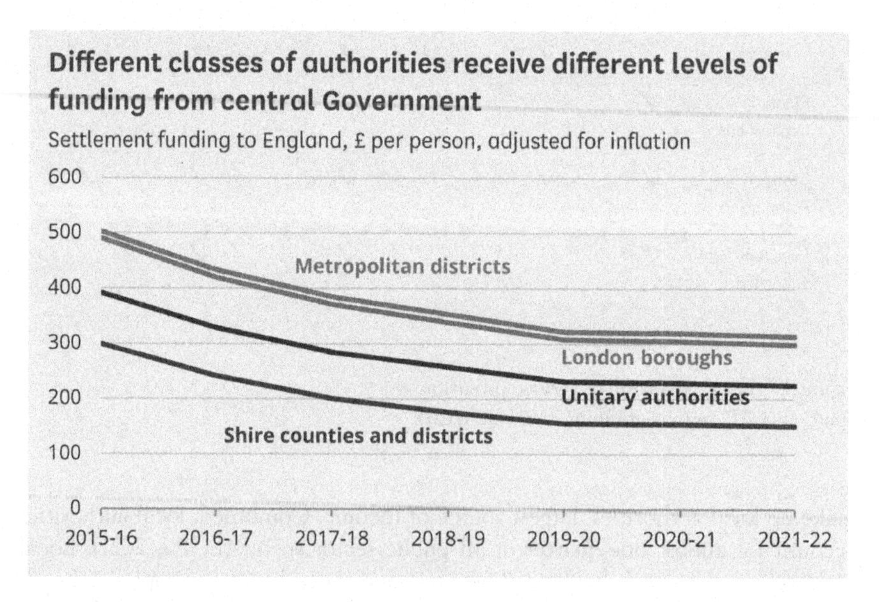

Figure 3.3 Local government funding, England
Source: MHCLG (2021).

Education services	32.6
Highways and transport services	3.8
Children Social Care	9.9
Adult Social Care	16.9
Public Health	3.2
Housing services (GFRA only)	1.8
Cultural and related services	2.2
Environmental and regulatory services	5.1
Planning and development services	1.4
Police services	12.2
Fire and rescue services	2.2
Central services	3.2
Other services	0.0
Total	**94.6**

Figure 3.4 Local authority spending by service type 2019–20, England (£billion)
Source: MHCLG, local authority revenue and expenditure and financing data, (2021).

Devolved local government system

In the past two decades, the UK has seen the devolution of powers to the Scottish Parliament, the national assemblies for Wales and Northern Ireland, and to London. Although certain fiscal powers have been devolved, the national Governments in Scotland, Wales and Northern Ireland are still dependent on central Government grants as their main source of funding for local government. In Scotland, Northern Ireland and Wales there is a single-tier system of local government providing all local authority services and they have a different structure for funding compared with England.

Since the 1970s the UK Treasury has used the Barnett Formula to calculate the annual block grants for the Scottish Government, Welsh Government and Northern Ireland Executive. It therefore determines the overall funding available for public services such as healthcare and education in the devolved nations. In 2019/20 the Barnett block grant amounted to £32bn in Scotland, £16bn in Wales and £12bn in Northern Ireland (before adjustments to account for tax devolution), reflecting differences in population size and the range of devolved public services in each nation (Institute for Government, April 2020).

Scotland

In Scotland, the Scottish Parliament Information Centre (SPICe) has published several briefings on local government financing. Here, funding for essential local government services is made up from the General Revenue Grant, non-domestic rates and council tax (see Figure 3.5).

In relation to expenditure, that of the Scottish local authorities exceeded their general funding in 2019–20 – a deficit that had to be funded from local authority reserves. However, according to the Scottish Government (2022), 'in 2020–21, total capital expenditure was £2, 604 million, a decrease of 31.4 per cent, or £1, 190 million, from 2019–20' (see Figure 3.6). Whilst Covid has had a significant impact on revenues, the decrease in capital expenditure also highlights the lack of alternative funding sources for devolved infrastructure investment.

The key sources of funding for Scotland's local governments derive mostly from grants and loans (Figure 3.7).

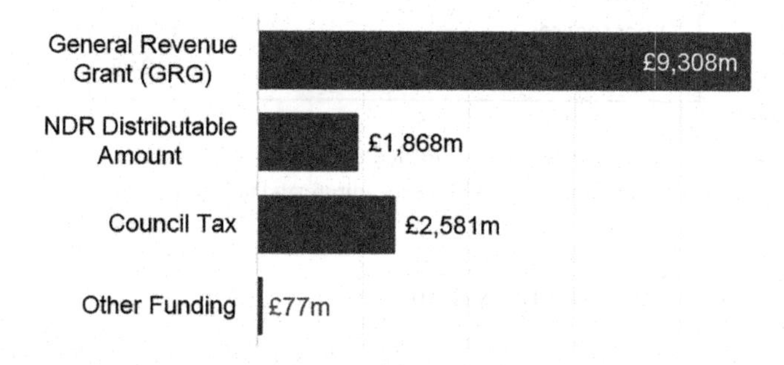

Figure 3.5 General funding in 2020–21 by source, Scotland (£million)
Source: *Scottish Local Government Finance Statistics 2020–21*.

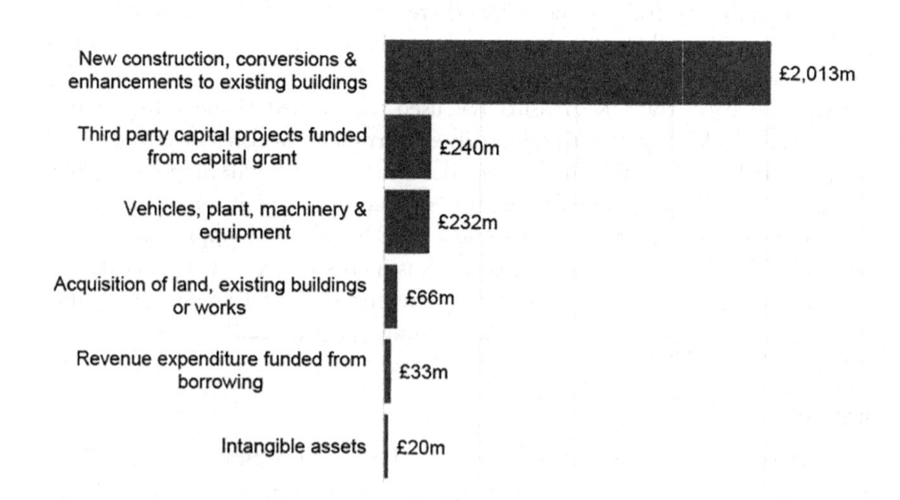

Figure 3.6 Capital expenditure in 2020–21 by expenditure type, Scotland (£million)
Source: *Scottish Local Government Finance Statistics 2020–21*.

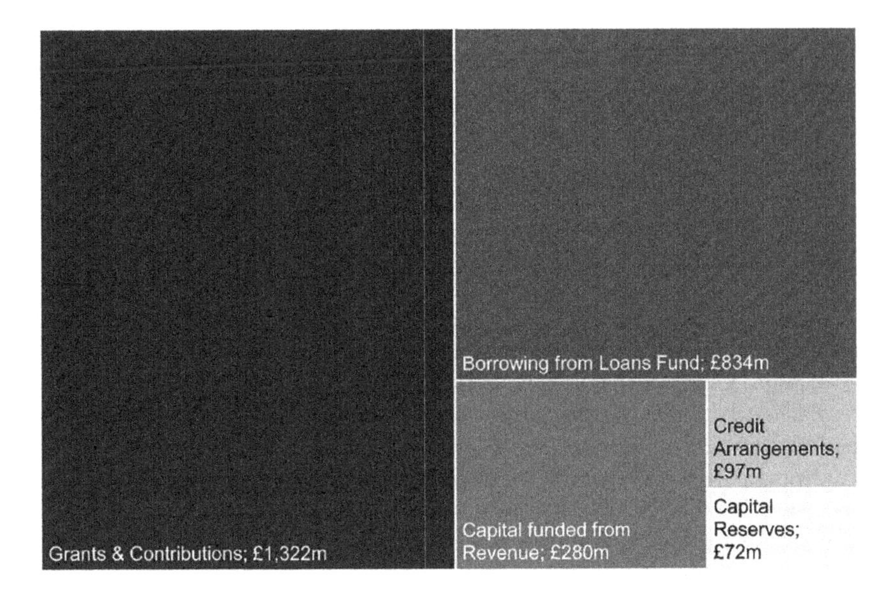

Figure 3.7 Financing of capital expenditure in 2020–21 by source, Scotland (£million)
Source: *Scottish Local Government Finance Statistics 2020–21.*

Wales

In Wales, local government receives around 80 per cent of its funding from the Welsh Assembly Government. Like Scotland, Welsh local government is financed through tax revenues and capital grants, and represents almost one third of the total Welsh budget. As can be seen in Figure 3.8, local government

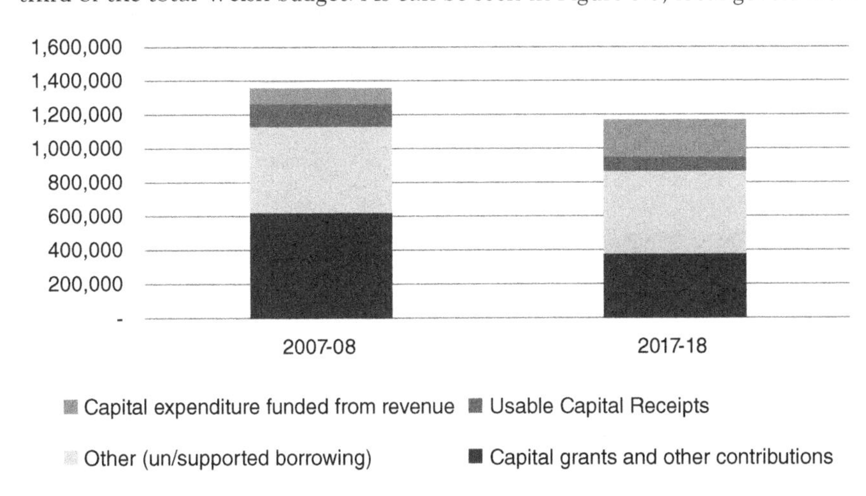

Figure 3.8 Local authority capital expenditure by funding source, 2007–08 to 2017–18, Wales
Source: StatsWales & HMT (GDP Deflators). Figure 1, WLGA (2019).

capital expenditure fell by 14 per cent over a ten-year period – from £1.36bn to £1.74bn – with funding from capital grants decreasing by 40 per cent in real terms. While externally supported borrowing has reduced by 49 per cent, unsupported borrowing – a term used to mean that funding has to be secured by the local authority – has risen by around 72 per cent in real terms over the same period.

Northern Ireland

The UK Government supported by the Northern Ireland Office and Westminster working groups performs a significant role in the delivery of infrastructure in Northern Ireland. Funding for regional infrastructure is secured through the block grant, City and Devolution deals, loans, and grants. As mentioned above, the Barnett Formula, introduced in the 1970s, is the mechanism used to adjust the level of funding for each devolved region accounting for changes in UK Government departmental spending objectives (Northern Ireland Office, 2019). It is calculated based on areas of the greatest need, and as such NI receives a greater share of public spending per head of population than other regions, totalling just under £22 billion in 2018–19 (Northern Ireland Office, 2019).

The NI Executive carries out most of the infrastructure spending in the entire region (Northern Ireland Office, 2019) and comprises nine departments that over the 8 year period to 31 March 2019 invested almost £10.6 billion on public infrastructure, of which the block grant contributed around 70 per cent of all infrastructure spending. Of the 11 local authorities, Belfast City Council (see Figure 3.9) is the largest, and in 2020/21 had a net budgeted expenditure of £167.1 million. Its largest source of funding is from district rates (70 per cent).

		2020/21			2019/20		
Service Expenditure	Notes	Gross Expenditure £	Gross Income £	Net Expenditure £	Gross Expenditure £	Gross Income £	Net Expenditure £
Strategic Policy & Resources	2	72,668,440	(31,739,096)	40,929,344	65,875,784	(13,685,316)	52,190,468
City Growth & Regeneration	2	31,313,623	(6,464,091)	24,849,532	32,164,175	(9,715,544)	22,448,631
People & Communities	2	115,946,789	(17,727,216)	98,219,573	120,253,125	(15,574,828)	104,678,297
Planning & Licensing	2	8,281,185	(5,219,447)	3,061,738	8,145,681	(6,360,679)	1,785,002
Cost of Services on Continuing Operations		228,210,037	(61,149,850)	167,060,187	226,438,765	(45,336,367)	181,102,398

Figure 3.9 Belfast City Council, Statement of Accounts, 2021
Source: Belfast City Council (2021).

Investment in local infrastructure

Central Government has cut overall local authority funding by about one-third since the global financial crisis of 2008. The Institute for Fiscal Studies (Amin-Smith et al., 2017) characterised local government finances in England by two major trends: large cuts to local authority budgets; and a shift from centralised redistribution of grant funding towards fiscal incentives for revenue growth. Furthermore, local authorities have tried to protect social care spending within their overall budgets. The protection of funding for social care has meant that investment in infrastructure, particularly housing and transport, has suffered. This sustained underinvestment in infrastructure has impacted negatively on regional economic growth; ergo a decrease in locally raised revenues and a greater dependency on central Government grants.

Hence, to deliver essential services, many local authorities are using their reserves. Local authorities can use their reserves to balance their budgets, but only for the short term and with an expectation that reserves will be replenished. Legally, local authorities must set a balanced budget and maintain an appropriate level of reserves. Over the past decade district authorities have increased reserves from around 50 per cent of annual spending to around 130 per cent, but county councils and unitaries (which both have responsibility for social care) have only increased their reserves from around 20 per cent of annual spending to less than 40 per cent (Amin-Smith et al., 2017).

Ultimately there is a funding crisis looming with regards the financial sustainability of local governments. The highest-profile case in recent years has been the ongoing budgeting problems of Northamptonshire County Council, which resulted in the Secretary of State appointing Commissioners in May 2018 to step in and take over key council functions. The measures that were introduced resulted in the council only being allowed to spend on those services that were a legal requirement. Sadly, this trend has continued. Croydon Council, which borrowed £545m to make a series of investments including a shopping centre and a hotel, saw the latter go into administration in June 2020, with the authority effectively declared itself bankrupt in November 2020. And in 2021 *The Guardian* reported that 7.4 per cent of councils in England were at high or acute risk of financial failure, and 94 per cent were planning to make cuts to balance their books (Butler, 2021).

The latest crisis facing local authorities is the Covid-19 pandemic. It has created extra pressures on local authorities who now risk cutting services and depleting their reserves. In response, the UK Government allocated £1.55 billion of un-ringfenced grants to local authorities to manage the immediate and long-term impacts of the pandemic. The government has also outlined details of a new 75 per cent guarantee scheme for irrecoverable council tax and business rates losses for 2020/21, worth an estimated £800 million (Ministry of Housing, Communities & Local Government, 2021).

In addition, the *2021 Spending Review* (SR21) pledges to make 'levelling up' a reality across Scotland, Wales and Northern Ireland by targeting action to meet local needs. SR21 provides an additional £8.7 billion per year on average to the devolved administrations over the SR21 period through the Barnett Formula on top of their annual £66 billion baseline. This will enable substantial investment into schools, housing, health and social care, and transport across Scotland, Wales and Northern Ireland (HM Treasury, 2021).

To summarise, in the UK there is a highly centralised system of taxation and expenditure. This has restricted the ability of local authorities to raise additional funds to invest in infrastructure. However, this system of centralised funding is changing due to the devolvement of powers to the regions and increasing complexities required to fund modern local governments. Essentially, local governments are funded from a taxation called business rates which are collected by local authorities and then redistributed by central Government on a 'needs-based' formula. As business rates is the key funding stream that allows local authorities to deliver their essential services, it is imperative that local authorities generate as much income from business rates as possible and in turn reinvest in their communities. Such reinvestment in local communities creates sustainable and resilient communities, which in turn create and sustain local government investment in local communities.

However, business rates are not a panacea for all the funding requirements of local government. High streets are changing; businesses are not surviving or growing in most high streets. Therefore, business rates collection is at risk. Before the catastrophic impact of Covid-19, the UK Government had already acknowledged the impact of failing high street retailers and the reduction in business rates by announcing a fundamental review of the business rates. This review has concluded with Government identifying a road map aimed at increasing business rates (HM Treasury, 2021).

In addition to business rates, local authorities also raise revenue from council tax on residential properties. Council tax is paid by households and is based on property values and can account for about 20 per cent of local authorities' total operating revenue. Given the pressures on spending, it is envisaged that nearly all local authorities plan to increase council tax on residents.

Local government capital spending

With a reduction in central Government funding, local authorities have sought to avail themselves of the following sources of alternative finance for capital expenditure:

a Borrowing
b Central Government grants
c Capital receipts from the sale of local authority assets
d Use of capital reserves
e Developer contributions for schemes agreed with local authorities

Local authority capital spending operates under different rules to revenue spending. Authority capital expenditure projects require 'that most local authority capital finance is obtained through borrowing. Authorities may also choose to finance capital projects via their reserves, or through various forms of joint venture with private sector bodies' (Sandford, 2021).

Local government borrowing

Under the Local Government Act (2003) local authorities can decide how much they borrow– this is known as the prudential borrowing scheme. The framework established by the Prudential Code should support local strategic planning, local asset management planning and proper option appraisal. The objectives of the Prudential Code are to ensure, within this clear framework, that the capital investment plans of local authorities are affordable, prudent and sustainable. Local authorities are required by regulation to follow the Prudential Code when carrying out their duties in England and Wales under Part 1 of the Local Government Act 2003, in Scotland under Part 7 of the Local Government in Scotland Act 2003, and in Northern Ireland under Part 1 of the Local Government Finance Act (Northern Ireland) 2011.

The prudential borrowing scheme represents a significant step away from the tight financial controls of central Government. The scheme is self-regulated, thus allowing local authorities to make long-term business plans. A strengthened Prudential Code, which aims to ensure local authorities' financial plans are affordable, prudent and sustainable, was published at the end of 2021.

In 2018–19 local authorities directly borrowed £93bn of capital finance (Sandford, 2021). For direct borrowing by local authorities to be approved the authority must demonstrate that the borrowing is both affordable, necessary, and repayable. In terms of repayable the local authority is required to identify how the capital investment could generate additional sources of revenue.

Direct borrowing for local government capital spending

The Public Works Loan Board (PWLB) was a statutory body of the UK Government that provided loans to public bodies from the National Loans Fund. Over 75 per cent of all local borrowing was via the PWLB and the interest rates for its loans were set by the HM Treasury. In October 2019, the Government increased the PWLB's lending rate from 0.8 per cent to 1.8 per cent over the gilt rate, in part to deter borrowing to fund risky investments. (The gilt rate is the rate at which the UK Government can borrow.) Many local authorities considered this increase in rates detrimental to their ability to deliver local capital investment requirements. In 2020, the PWLB was abolished as a statutory organisation and its functions were allocated to HM Treasury, where they are discharged through the UK Debt Management Office.

Local government borrowing – bank finance

Local authorities can also avail themselves of commercial bank lending. Local authority commercial lending is characterised by long-term Lender Option Borrower Option (LOBO) loans. These often start with a fixed interest rate, which is usually more attractive than the (former) PWLB offer. However, the lender can re-fix the interest rate periodically over the life of the loan, which has led to criticism of this financial instrument as often the local authority borrower is forced either to accept the interest rate change or to repay the loan. This can often result in the local authority having to make spending cuts to make the loan repayment. Research by Local Government (2021) shows that 95 per cent of outstanding LOBO debt is now owed to European Banks, with councils projected to pay at least £14bn in interest payments until the end of the loan period. That said, local authorities are typically only modestly exposed to LOBOs and these loans make up only a small proportion of overall debt. Also, in recent years, there has been a trend for local authorities to exit from these types of financing arrangements which are considered 'toxic, expensive and risky' (Local Government, 2021).

Loans from multilateral lending institutions such as the European Investment Bank (EIB) have also been important sources of cheap long-term finance for local authorities. Following Brexit, it is also possible that the EIB could continue to lend to the UK as a non-EU member.

Local government borrowing – bond finance

Local authorities can make use of bond finance for capital expenditure. Bond finance is typically provided by large institutional investors such as pension funds and insurance companies. The local authority or municipal bond market has yet to be significantly developed, and the Local Government Association (LGA) has backed the setting up of a Municipal Bonds Agency (MBA), which would be co-owned by a number of UK local authorities, to help develop this market in the UK. It is hoped that the full development of a municipal bond market would stimulate investor interest and increase the long-term lending of bonds to local authorities.

However, a key issue for institutional investors in the issuing of bonds is a credit risk assessment of the local authority they are lending to. Whilst loans to Government are considered low risk, some local authorities are considered high risk. This is further compounded by the lack of published credit ratings by local authorities. Another issue for institutional investments is the bureaucracy of local government. This layer of complexity leads to uncertainties and difficulties in satisfying the requirements for transparency and accountability set by the local authorities.

Local government borrowing – pension funds

Pension funds have traditionally not sought to invest in the construction and development phase of infrastructure, but have instead looked to invest after construction. As projects set out in the National Infrastructure Delivery Plan start to be delivered, and with a Government that is keen to encourage long-term investment in infrastructure, there may be more investment opportunities for local authority pension funds in the future.

Although local authority pension funds are also a source of finance for local authorities, there are a plethora of different ones across the UK. This fact alone makes investment using local authority pension funds complicated. However, Manchester City Council was able to utilise local authority pension funds to deliver a mixed tenure housing scheme of over 200 homes. The project was delivered in collaboration with Manchester City Council, Homes England and the Greater Manchester Pension Fund.

Case Study – Greater Manchester Pension Fund

The Greater Manchester Pension Fund (GMPF) is the largest local Government pension fund in the UK. In 2012, the GMPF signed an agreement between Manchester City Council and the Homes and Community Agency (replaced with Homes England) to invest in new-build affordable homes. The Greater Manchester Property Venture Fund (GMPVF) has an allocation of up to £750 million and is directly involved in a number of construction and development projects (GMPF, 2021).

The project at a glance:

Dates: 2012–

Cost: £750m to date

Project goals:

Financing of a range of housing schemes including

- 240 new homes on land provided by the council.
- 2 towers, 37 and 17 storey, housing 683 private rented apartments on Manchester's Oxford Road.
- A 164-flat development at Pomona Island along the Manchester Ship Canal
- The GMPF is also supporting the Greater Manchester Homes Partnership, which was set up in 2017 by the combined authority's elected mayor Andy Burnham to tackle rough sleeping in the city.

Project funding:

The investment model was the first initiative of its kind in the country and provided a fresh way to invest in new-build affordable homes. The three parties have signed a memorandum of understanding under which the council will provide most of the land for the sites, the HCA will provide one of the sites and the GMPF will deliver the funding.

Project benefits:

The GMPF has delivered housing investments in areas in need of property and provides decent, affordable homes for communities. The housing projects have also realised additional benefits in local communities included providing investments to local Small to Medium Enterprises (SMEs), Social Impact Bonds (SIBs) and renewable energy infrastructure.

Tax increment financing (TIF)

Tax increment financing (TIF) is a financial instrument considered by local authorities to secure funding for capital projects. The project is part-funded by the projected future increase in business rates income associated with the new infrastructure. TIF is used as a subsidy for capital infrastructure investment. According to the Scottish Futures Trust (2021):

> The use of TIF is normally predicated on a 'but for' test i.e. that but for TIF, the anticipated outcomes from a regeneration and economic perspective would not occur or not occur in the time frames which TIF would enable.
>
> A TIF project must therefore demonstrate that the enabling infrastructure will generate additional public sector revenues to repay the financing requirements of the enabling infrastructure.

Central Government (capital) grants

Capital grant funding for local authority capital investment programmes was the main source of additional funding for key areas such as schools, roads and housing. Grants were provided to local authorities on the basis of an established and complex formula of specific needs. Before Brexit the grants were provided by the EU through various funds. In 2017, the Local Government Association published *Beyond Brexit: Future of Funding Currently Sourced from the EU*, which identified around €10.5bn of EU funding that will need to be replaced after Brexit. In response, Government announced a UK Shared Prosperity Fund in 2022 to replace EU grant funds (Brien, 2022). It is expected to be less bureaucratic and more targeted on schemes that work to improve the local economy.

Capital receipts from local authority asset disposal

Local authorities can also use capital receipts from the disposal of their assets to meet the costs of capital investments. In 2016, the then ministry for Housing, Communities and Local Government published guidance for local authority asset disposal 'to support more efficient and sustainable local services' (Department for Communities and Local Government, 2016). Local authorities can also provide land as equity in a development project and share in the profits. Councils are required to publish details of their land and buildings to comply with the Transparency Code (2015), which is a key element of public trust and confidence in local government officials.

Capital reserves

Local authorities have the capacity to utilise their capital reserves for capital expenditure. However, the local authority must maintain appropriate levels of general reserve funding for the day to day running of the municipal, which can limit the availability of funds for infrastructure investment (Sandford, 2021). As an example, Transport For London has successfully generated ringfenced capital reserves through the introduction of the Congestion Charge, not only does this reduce city centre traffic, but it also provides significant funding for transport infrastructure around the municipal (TFL, 2014).

Developer contribution

The Community Infrastructure Levy (CIL) is another financial instrument that local authorities can use to secure additional funding to support capital developments. It was introduced by the Government in 2014 and last updated in 2022 by the Department for Levelling up, Housing and Communities (DLUHC) (DLUHC & MHCLG, 2022). The CIL is a charge which can be levied by local authorities on new development in their area. Latest guidance from the DLUHC on the CIL states that the:

- The levy only applies in areas where a local authority has consulted on, and approved, a charging schedule which sets out its levy rates and has published the schedule on its website.
- Most new development which creates net additional floor space of 100 square metres or more, or creates a new dwelling, is potentially liable for the levy.
- Some developments may be eligible for relief or exemption from the levy. This includes residential annexes and extensions, and houses and flats which are built by 'self-builders'. There are strict criteria that must be met, and procedures that must be followed, to obtain the relief or exemption.

Local authorities are required to makes sure they are clear about the infrastructure needs associated with specific developments, and what developers are expected to pay for supporting infrastructure.

Commercial developers can also make contributions to local infrastructure investment through agreements under section 38 of the Highways Act 1980. These require developers to invest in road works associated with a development. The works must be delivered to a particular standard and timescale. The local authority then agrees to maintain the street at public expense once the work is carried out to its satisfaction.

City and Regional Growth Deals

The Government introduced City Deals in 2012 (HM Government, 2012) and Local Growth Deals (in 2014) to 'provide funds to local enterprise partnerships or LEPs (partnerships between local authorities and businesses) for projects that benefit the local area and economy'. This further enhances the roles of local authorities to deliver local and regional capital investment.

City Deal agreements have now been subsumed into Devolution Deals. The Cities and Local Government Devolution Act 2016 (see Figure 3.10) allows combined authorities to take on more powers and functions, but there must be an elected mayor to take on these responsibilities. The powers given to elected mayors representing combined authorities vary between different areas. The mayors are expected to provide strong voices for the areas they represent, and further devolved powers may be acquired over time. City or Regional Growth Deals (as they are called in Scotland) are a negotiated package of funding for cities and regions across the UK. The packages are negotiated between central, devolved and local government. The funding packages are aimed at channelling additional investment, creating new jobs and enhancing economic growth.

1 The Act is intended to support delivery of the Government's manifesto commitment to "devolve powers and budgets to boost local growth in England", in particular to "devolve far-reaching powers over economic development, transport and social care to large cities which choose to have elected mayors" and "legislate to deliver the historic deal for Greater Manchester". The Act takes forward a number of reforms which are intended to allow for the implementation of devolution agreements with combined authority areas and with other areas. It is enabling legislation which provides a legislative framework which can be applied flexibly to different areas by secondary legislation. It will enable secondary legislation to:

- enable any public authority function relating to an area to be conferred on a county council or district council;
- provide for streamlined local governance as agreed by councils;
- provide for fast track structural and boundary changes to non-unitary local authority areas where at least one local authority asks for this;
- enable any public authority function relating to an area to be conferred on a combined authority;
- confer any local government function on a combined authority (these are currently limited to economic development, regeneration, and transport);
- provide for an elected mayor for a combined authority's area who would exercise specified functions individually and chair the authority; and
- provide for the possibility for the mayor additionally to undertake the functions of a Police and Crime Commissioner for the combined authority area (in place of the Police and Crime Commissioner).

Figure 3.10 An overview of the Local Government Devolution Act 2016
Source: HM Government (2016).

Case Study – Liverpool City Deal

In 2012, Liverpool was the first city to be awarded a City Deal. Specific to the Liverpool City Deal was the desire to enhance economic opportunities and in doing so realise the ambitions of Liverpool to become a world class international city. The City Deal further builds on the private and public partnership that exists between the Local Enterprise Partnership (LEP), City Region Leaders and the Mayor of Liverpool. 'Aside from the localism approach of central government in allowing each region to design and create their own specific investment opportunities, another key driver behind establishing the City Deal was to encourage collaboration and cooperation between private and public sector actors' (O'Brien & Pike, 2019). It focuses on transport and skills projects which will support the city region's ambitions to create a freight and logistics hub serving an expanded Port of Liverpool.

The project at a glance:

Dates: 2012–

Cost: £900 million

Project goals:

Infrastructure schemes include:

- Liverpool and Wirral Waters (planned £10bn private investment);
- International Trade Centre (planned £200m private investment);
- New Deep Water Port on the Mersey (planned £300m private investment);
- Expansion of Daresbury Science and Innovation Park (planned £600m private and public investment);
- Mersey Gateway Bridge (planned £450m PFI investment);
- Expansion of the Mersey Multimodal Gateway (3MG) (£100m private investment);
- New Royal Liverpool University Hospital and Bio Campus (£500m PFI investment).

Project funding:

As O'Brien and Pike (2019) suggest,

> City Deal funding is a relatively new method of devolved governance which allows local regions to lead their own regeneration efforts, by

negotiating an agreement between local region and central government in a bid to be awarded financing to further infrastructure investments with the over-arching aim of boosting economic performance.

In addition to the £900 million devolution deal, the Liverpool City Region Combined Authority has secured more than half a billion pounds in additional funding from government, including:

- £173 million in Transforming Cities Funding which are for modernising and improving the way we travel in and around the city region.
- £8 million in extra funding to tackle homelessness and rough sleeping.
- A share of £300 million to connect Liverpool to the HS2 network.
- £232.3m from the Government's Local Growth Fund.
- £35m of new funding confirmed for 2015/16.
- £153.2m of funding over the years to 2020/21.

As well as the Government's investment in the Liverpool City Deal, the private sector is expected to invest in the region.

Project benefits:

Steve Rotheram, Metro Mayor of the Liverpool City Region, announced that 'By taking power, money and decision making out of Westminster and putting it in the hands of local people, we can build the fairer, greener future we want for our region'.

By 2021 Government will have further invested over £12bn through the Local Growth Fund to deliver jobs and investment for the Liverpool City Region economy. The Region will benefit from more than 18, 000 new jobs and £225m of additional investment as a result of £313m worth of Local Growth Fund investments. In fact, the investments made in the first three years will deliver 9, 000 jobs and 5, 500 apprenticeships for local people.

Projects backed by the Combined Authority include the redevelopment of Kirkby Town Centre, Liverpool's City Centre Connectivity scheme and the construction of a new train station at Maghull North – the first new station on the Merseyrail network in a generation.

Summary

The funding of local governments is experiencing a time of critical change. The traditional reliance of local authorities on the central Government's reallocation of business rates has started to decrease. This move away from traditional sources of funding has been necessitated by the decrease in central Government funding and the increase in other forms of direct borrowing. This move has been supported by successful Governments who have pushed the localism agenda and levelling up agenda throughout the UK.

Legislation has also been enacted to provide local authorities more autonomous fiscal powers. There is still concern that many local authorities will not manage to 'balance their books' and it is uncertain what support they may receive from central Government. Long-term financing for local infrastructure projects remains difficult due a range of factors including the credit rating of local councils, the bureaucracy of local government and the lack of interested investors. However never more than now do local authorities need to invest in infrastructure to ensure the viability of their towns and cities, without which the economic outlook for the entire region will be bleak.

References

Amin-Smith, Neil et al. (2017, September). "The local vantage: How views of local government finance vary across councils". IFS Report, No. R131. Available at: http://dx.doi.org/10.1920/re.ifs.2017.0131

Belfast City Council (2021, March). *Statement of Accounts 2021*. Available at: https://minutes.belfastcity.gov.uk/documents/s93853/Statement%20of%20Accounts%202021.pdf

Brien, P. (2022, April). *The UK Shared Prosperity Fund*. Research briefing. Available at: https://researchbriefings.files.parliament.uk/documents/CBP-8527/CBP-8527.pdf

Butler, P. (2021). "Swingeing cuts on cards as councils in England face funding crisis, watchdog warns", *The Guardian*, 10 March 2021. Available at: https://www.theguardian.com/society/2021/mar/10/swingeing-cuts-on-cards-as-councils-in-england-face-funding-crisis-watchdog-warns

CIPFA (2021). *The Prudential Code for Capital Finance in Local Authorities* (London: CIPFA).

Department for Communities and Local Government (2015, February). *Local Government Transparency Code 2015*. Available at: https://assets.publishing.service.gov.uk/government/uploads/system/uploads/attachment_data/file/408386/150227_PUBLICATION_Final_LGTC_2015.pdf

Department for Communities and Local Government (2016). *Local Authority Assets Disposal Guidance*. Available at: https://assets.publishing.service.gov.uk/government/uploads/system/uploads/attachment_data/file/508307/160316_Land_disposal_guidance.pdf

DLUHC & MHCLG (2022, April). *Community Infrastructure Levy*. Available at: https://www.gov.uk/guidance/community-infrastructure-levy

Greater Manchester Pension Fund (2021). *GMPF 2021 Annual Report*. Available at: https://www.gmpf.org.uk/Kentico12_Admin/GMPF/media/About/documents/GMPF-Annual-Report-2021.pdf

HM Government (2012). "Liverpool City Deal". Available at: https://www.gov.uk/government/publications/city-deal-liverpool

HM Government (2012, July). *Unlocking Growth in Cities: City Deals – Wave 1*. Available at: https://assets.publishing.service.gov.uk/government/uploads/system/uploads/attachment_data/file/7524/Guide-to-City-Deals-wave-1.pdf

Institute for Government (2020, April). "Barnett Formula". Available at: https://www.instituteforgovernment.org.uk/explainers/barnett-formula

Liverpool City Region (2021, June). "£313m Local Growth Fund fully invested to deliver jobs and investment for the Liverpool City Region economy". Available at: https://www.liverpoolcityregion-ca.gov.uk/313m-local-growth-fund-fully-invested-to-deliver-jobs-and-investment-for-the-lcr-economy/

Liverpool City Region (n.d.). "Growing the Liverpool City Region economy". Available at: https://www.liverpoolcityregion-ca.gov.uk/growing-our-economy/

Local Government Association (2017, July). "Beyond Brexit: Future of funding currently sourced from the EU". Available at: https://www.local.gov.uk/topics/european-and-international/beyond-brexit-future-funding-currently-sourced-eu

Local Government Association (2021). "What is local government?". Available at: https://www.local.gov.uk/about/what-local-government

Local Government Association (2021), "Transparency code". Available at: https://www.local.gov.uk/our-support/guidance-and-resources/data-and-transparency/local-government-transparency-code; https://assets.publishing.service.gov.uk/government/uploads/system/uploads/attachment_data/file/408386/150227_PUBLICATION_Final_LGTC_2015.pdf

Local Government Information Unit (2021). "Local government facts and figures: England". Available at: https://lgiu.org/local-government-facts-and-figures-england/#section-5

Local Gov (2021, October). "More councils urged to exit toxic LOBO", 27 October 2021. Available at: https://www.localgov.co.uk/More-councils-urged-to-exit-toxic-LOBO-debt/53167

MHCLG & DLUHC (2021). "Final local government finance settlement: England, 2021 to 2022". Collection. Available at: https://www.gov.uk/government/collections/final-local-government-finance-settlement-england-2021-to-2022

Ministry of Housing, Communities & Local Government (2021, July). December 2020: "COVID-19 funding for local government in 2021 to 2022 – consultative policy paper". Guidance. 15 July. Available at: https://www.gov.uk/government/publications/covid-19-emergency-funding-for-local-government/covid-19-funding-for-local-government-in-2021-22-consultative-policy-paper

Northern Ireland Office (2019). *Annual Report and Accounts 2018–19*. HC 52. Available at: https://assets.publishing.service.gov.uk/government/uploads/system/uploads/attachment_data/file/841932/NIO_Annual_Report_and_Account_for_laying_on_2410191_-_certified_by_C_AG_on_231019__2_.pdf

O'Brien, P. and Pike, A. (2019). "'Deal or no deal?' Governing urban infrastructure funding and financing in the UK City Deals", *Urban Studies*, vol. 56, no. 7, pp. 1448–1476.

Sandford, M. (2021). Local Government in England: Capital Finance, House of Commons Library. Available at: https://researchbriefings.files.parliament.uk/documents/SN05797/SN05797.pdf

Scottish Futures Trust (2021). "Tax Increment Finance (TIF)". Available at: https://www.scottishfuturestrust.org.uk/page/tax-incremental-financing

Scottish Government (2021). *Scottish Local Government Finance Statistics (SLGFS) 2020–21*. Collection. Available at: https://www.gov.scot/publications/scottish-local-government-finance-statistics-slgfs-2020-21/pages/4/

Transport for London (2014). "Congestion Charge factsheet". Available at: https://content.tfl.gov.uk/congestion-charge-factsheet.pdf.pdf

Wales Fiscal Analysis (2021). "Local government & the Welsh budget (2021)". Available at: https://www.cardiff.ac.uk/__data/assets/pdf_file/0009/2513619/lgf_outlook_2021_7.pdf

Welsh Local Government Association (2019, April). "WG's capital funding sources – MIM". 29 April. Available at: https://www.wlga.wales/SharedFiles/Download.aspx?pageid=62&fileid=2404&mid=665

4 Infrastructure procurement

Introduction

This chapter explores infrastructure procurement models, discusses their inherent characteristics and identifies how risk is managed in the procurement process. The aim is to provide readers with a clear understanding of how these models operate, transfer risk and deliver essential infrastructure investment that benefits all in society. These models are designed to harness the private sector investment and expertise required to deliver and manage infrastructure development. This section provides an overview of the relevant procurement frameworks, including new regional infrastructure procurement models, together with a lessons' learned section on the risks and failings of PFI and PF2.

Learning outcomes

In this chapter you will learn about:

1 Generic infrastructure procurement models, including delivery consortia, development partners, alliancing and frameworks.
2 Issues surrounding the procurement of private sector partners in the delivery of infrastructure projects.
3 Infrastructure procurement frameworks specific to Scotland, Wales, and Northern Ireland.

The infrastructure procurement decision

Government's decision to invest in new public infrastructure or to deliver an essential public service or maintain a public asset, is both strategic and fundamental to the wellbeing of all in society. The importance of infrastructure investment can be identified as a measure of direct economic contribution and the delivery of social benefits to local communities. Significantly, these local economic and community benefits contribute nationally through job creation and growth, and to the overall well-being and health of citizens, to create vibrant, attractive and sustainable communities.

DOI: 10.1201/9781003171805-4

Despite the national importance of infrastructure investment in creating such communities, governments cannot entirely deliver society's infrastructure needs. Instead, they must determine and plan infrastructure needs, and seek to procure infrastructure using alternative public private procurement models. In the UK, a mix of procurement models have been adopted to deliver infrastructure projects, which include:

1 Central Government stakeholder
2 Local government client stakeholder
3 Private partner stakeholders

Delivery stakeholder partner and consortia

In the UK, the delivery partner/delivery consortia procurement model adopted is that of Public Private Partnerships (PPPs). PPPs are used when the public sector requires a private sector partner to design, build, finance, maintain and operate a public asset. HM Treasury (2013) has suggested that infrastructure projects requiring major investment should consider the PPP model if:

- funding and finance from private sector are required;
- the infrastructure project meets society's long-term needs;
- there is adequate and transparent allocation of risk between public and private sectors;
- the whole life cost of the infrastructure can be ascertained;
- the project is technically advanced, ergo requiring specialist input from the private sector; or
- the cost of construction is more than £50m.

Development/delivery partners

Development/delivery partners in infrastructure are best considered in terms of the approach taken to deliver the London Olympic Games in 2012 and the Crossrail programme. Delivery partners are organisations contracted to deliver a specific programme and, as such, do not have permanent status. Due to the focused and specific nature of the infrastructure programme, this procurement model is by nature a highly flexible, integrated and collaborative approach, capable of delivering high levels of project success. It is, however, considered essential that the procuring authority should seek to maintain a degree of risk and control over the delivery consortia model.

Alliancing

Alliancing infrastructure procurement is considered by the UK Government to be appropriate for both low-to-medium value complex infrastructure projects and medium-to-high value complex infrastructure projects. Each of the alliance partners will work collaboratively within an integrated expert supply chain to develop and manufacture development solutions. Unlike partnering frameworks, the alliance framework is structured to facilitate multiple contracts between the procuring authority and individual alliance partners. For example, the Anglian Water Business Plan 2020–2025 (Anglian Water, 2018) advocates the successful use of an alliance approach to deliver a major project due to a considerable number of project benefits.

Frameworks – policy and legislation

Procuring of infrastructure projects using public private consortiums or frameworks are fundamentally based on the allocation of risk and fairness of risk for each partner in the framework. In allocating risk to each partner, authorities procuring a framework partner can adopt a collaborative procurement strategy. The features of an effective framework are highlighted in the National Construction Category Strategy for Local Government 2020 (Local Government Association, 2020) which states that 'effective frameworks can offer distinct benefits over traditional procurement for projects by facilitating a more integrated solution based on continuing and closer relationships with a limited number of suppliers'.

Furthermore, infrastructure procuring authorities are no longer required to comply with EU rules pertaining to the Public Sector Procurement Directive (2004/18/EC), which set out the rules for the procurement of goods, services and works above certain threshold values and was embodied within the Public Contracts Regulations 2006 (PCR). Instead, from 2021, the UK has been following the WTO GPA (Government Procurement Agreement) – a voluntary plurilateral agreement intended to promote transparency and fairness (WTO, 2021). As other countries following the GPA include both EU member states and non-EU member states, the obligations and rights of GPA-abiding parties are in sync with those of the OJEU.

In response to the withdrawal from EU procurement rules, the UK Cabinet Office (2020), stipulated in their Procurement Policy Note 08/20 (PPN):

> whilst the framework and principles underlying the public procurement regime (the procurement procedures, financial thresholds, etc.) will not substantially change, contracting authorities will be required to publish public procurement notices for new procurements to the new UK e-notification service, Find a Tender.

Thus, public procurement must now follow the Find a Tender protocol from the end of the Brexit transition period. This PPN applies to all contracting authorities – which it refers to as Central Government Departments, Executive Agencies, Non Departmental Public Bodies, wider public sector, local authorities, NHS bodies, and utilities – 'in respect of procurements regulated by the Public Contracts Regulations 2015, the Utilities Contracts Regulations 2016, the Concession Contracts

Regulations 2016 and The Defence and Security Public Contracts Regulations 2011'. It gives the following guidance on the adoption of Find a Tender:

1. Procurements on OJEU/TED that were commenced prior to the end of the Transition Period must be concluded on OJEU/TED.
2. New procurements commenced after the end of the Transition Period must be advertised on Find a Tender.
3. Requirements to advertise on Contracts Finder, MOD Defence Contracts Online, Public Contracts Scotland, Sell2Wales and eTendersNI remain unchanged.

Source: Cabinet Office (2020)

However, the Government recognises that there is no 'one size fits all' procurement model for the delivery and management of infrastructure projects. Instead, they have identified several conventional attributes that should be embedded in bespoke infrastructure procurement methods. The infrastructure procurement attributes as identified by the HM Treasury & UK Infrastructure publication on 'improving delivery capability' are:

- collaborative working;
- appropriate risk allocation;
- incentivisation of the supply chain at 1st- and 2nd-tier level; and
- approaches to supply chain performance management.

The Government recognises that the most important attribute of infrastructure procurement is the approach taken to managing the risks associated with delivering infrastructure projects. The next subsection presents a comprehensive review of infrastructure procurement models and identifies how risk is managed to best ensure the maximum economic and social benefits are derived from infrastructure investment.

Frameworks and risks – in transition

Following a period of underinvestment in infrastructure, the UK Government determined that private sector involvement was necessary and advantageous to reverse the trends in infrastructure investment. As a result, Public–Private Partnerships (PPPs) were introduced as procurement models that could encapsulate the provision of infrastructure projects through a joint public and private framework agreement.

PPPs were introduced in the UK by the Conservative Government in 1992 and extolled as a procurement method that could deliver and finance infrastructure projects. The PFI (Private Finance Initiative) was seen as 'an initiative to increase the amount of private finance being used to invest in what had previously been publicly sponsored infrastructure' (Hellowell, 2008). According to NAO (2018) figures, there were at that time over 700 PPP contracts in operation, constituting over £60 billion (bn). This equates to annual charges of circa over £10.3bn and a total of over £199bn into the 2040s if no further deals were agreed.

The procurement of infrastructure using PPP fundamentally differs from all other methods of procurement in that the private sector funds and finances the asset through its whole life. This requires the private sector partners to play a pivotal role in maintaining and managing the public asset for a period of up to 25–30 years, or its useful life. Although the public sector client pays for the asset, they will make smaller capital payments over the PPP agreement duration, making it an attractive alternative to entirely funding the infrastructure project.

Comparatively, more modern definitions of PPP establish it to be a legally binding contract or other mechanism where partners agree to share responsibilities related to the implementation and/or operation and management of an infrastructure project. Van Ham and Koppenjan's (2001) definition is narrower and more helpful, describing PPPs as the act of cooperation between both public and private sector actors, in which products and services are jointly developed and the risks, costs and resources needed to deliver these are shared. Garvin and Bosso (2008) furthered this to focus on the utilisation of private finance (which is put at risk), management and operating services. Hodge and Greve (2007) have a similar narrow definition to that of Gavin and Bosso, but include the requirement for new accountability and governance requirements to appropriately focus the parties on the successful delivery of infrastructure.

These projects are procured by adopting the Private Finance Initiative (PFI) model and latterly adopting the Private Finance 2 (PF2) model. Until 2010, the primary PPP procurement model was PFI. However, in 2012, HM Treasury produced a report identifying the inherent failings of this framework concerning the transfer of risk and the resultant value. These major criticisms ultimately resulted in the demise of the PFI and led to the creation and implementation of a new suite of PPP models, presented below in the section on 'National infrastructure procurement models'.

Risks and failings of PFI

The UK Government committed to the PFI procurement model from the mid-1990s to 2010. The heavy adoption of the PFI model was based on the Government's conviction that the PFI framework could best deliver value for money through an elaborate public sector–private sector risk sharing framework.

However, according to the HM Treasury select committee report on PFI (2012), the PFI framework was often deficient and ineffective, and thus incapable of delivering value for money. The report found that the over-apportioning of risks from the public sector to the private sector had created huge windfall profits for the tier one private sector at the expense of the public purse. Conversely, the report also found that the private sector delivery consortia were often unable to appreciate, plan and mitigate PFI project risks. This resulted in the private sector incurring financial losses. Other failings of PFI included the very high cost of tendering for

PFI frameworks, risk of financial losses for the lower tier private sector consortia and the extensive procurement process.

Collectively, these criticisms and failings deterred the private sector, including investors, from engaging in PFI frameworks. One of the key failings of PFI projects was considered to be the over-eagerness of Government to favour PFI frameworks often when alternative and more traditional procurement methods could have resulted in greater value for money. These criticisms led to the discontinuation of the PFI framework.

Following the HM Treasury report, Government is now both committed and determined to engage effectively with the private sector for the delivery and funding of public infrastructure. This commitment has manifested in the introduction of new measures aimed at reducing risks for both parties and therefore improving a project's chances of success. These new measures have led to the development and roll-out of new infrastructure procurement frameworks across the UK, which have been encapsulated in HM Treasury paper 'Analysis of the National Infrastructure and Construction Pipeline 2021'.

National infrastructure procurement models

The 2008 global financial crisis and the multiple failings of the PFI framework deterred private sector investment in public infrastructure projects. The Conservative–Liberal Democrat coalition Government therefore paved the way for a raft of new infrastructure procurement frameworks. Key to addressing the multiple failings of PFI was the critical examination of the inherent risks of delivery consortia PPP frameworks and resolving these to mitigate risks for both the public purse and private sector investors.

Realigning risk in infrastructure procurement frameworks facilitated the public sector client being redefined as the key authority stakeholder. Furthermore, specifically within PPP frameworks, the public authority could now formally adopt the name 'sponsor'. As the sponsor of infrastructure projects, the public sector could co-invest in the funding of infrastructure projects, reducing the financial risks for the private sector. The reduction of private sector financial risks created a more attractive framework for all stakeholders and provided government with essential commitment from the private sector.

Acknowledging the inherent weaknesses and risks of the PFI framework, and the failure to deliver value for money in PPP projects, necessitated a new approach to delivery consortia – essentially a new PPP framework approach that apportioned risks robustly and equitably. This new approach was the Private Finance 2 (PF2) framework. PF2 was announced in December 2012 by the Government in a document entitled *A New Approach to Public Private Partnership*.

Private Finance 2 (PF2)

In 2013, the coalition Government launched PF2 to replace PFI. Its introduction affirmed the Government's commitment to continued collaboration with the private sector for the delivery and funding of public assets – a commitment that is rooted in their belief that the PPP model can deliver value for money. Specifically, that value for money can be harnessed and acquired through private sector innovation and skills, and ultimately the transference of risk from the public sector to the private sector.

A fundamental shift in allocation of risks between the PFI and PF2 models was the introduction of the public sector as an investor of finance. Acting in this new role as co-investor, the authority can directly invest in the project and therefore take on greater responsibility and risk. This transfer of pre-determined risks from the private partners for the financial aspects of the project to the public sector was one of many new risk reduction features. Other risk reduction features and framework improvements embedded in PF2 included:

- restricted ability of the private sector to make windfall gains;
- greater distribution in the allocation of risk between partners;
- the establishment of a board for each PF2 project that reports annually on profits made by the private sector consortia; and
- soft services contracts would now operate for a reduced period of 5 years.

The PF2 process is shown in Figure 4.1.

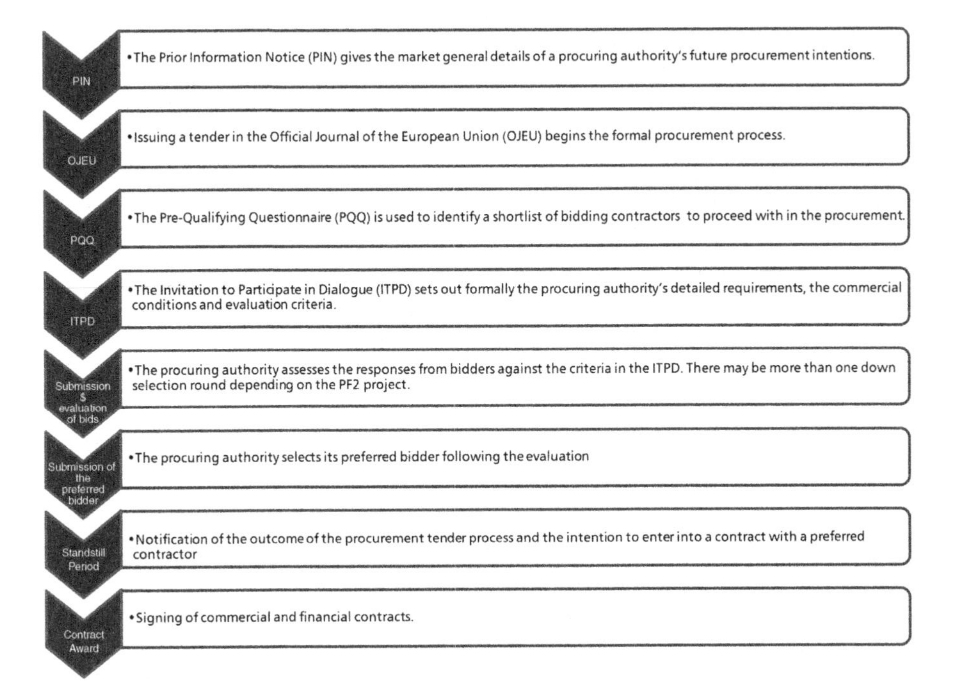

Figure 4.1 The PF2 process
Source: HM Treasury (2012).

PF2 framework and risks

As a direct consequence of the failings of PFI, the PF2 model had the following improved features:

1 Streamlining the procurement process

The more protracted procurement timeframe (~ 3–4 years) was a critical weakness of PFI and deterred the private sector from investing in PPP schemes. In PF2 projects the procurement phase was condensed to 18 months, with a commitment to reduce the procurement timetable further. To ensure the cap on the procurement timetable was not exceeded, PPP projects with a procurement phase exceeding 18 months would no longer be approved by HM Treasury unless an exemption had been granted by the chief secretary.

Additional Treasury checks were introduced prior to projects going to the tender stage, thereby ensuring the robustness of the project's suitability for PPP procurement. These checks were put in place to further reduce the risks arising from projects not delivering value for money. PF2 also introduced the Government's lean sourcing principles, designed to deliver value for money on infrastructure projects through the introduction of a checklist approach.

Collectively, the procurement framework changes mitigated the risks of failing or inappropriate adoption of PPP. Furthermore, the shortened and streamlined procurement process reduced tendering costs, improved the time taken to get the project to the post-contract and construction phase, and bolstered private sector interest in PPP projects, thus enhancing both competition and quality.

2 Standardisation of contract documentation

To compliment the streamlined procurement process and to further simplify infrastructure procurement, PF2 provided a comprehensive suite of contract documentation. The standardisation of contract documentation aimed to address ambiguities in the PPP framework by explicitly and expressly identifying the roles and obligations of each stakeholder and identifying how the risks of the infrastructure project were to be shared. An example of this risk sharing was the obligation of both public and private sector stakeholders to share the insurance premiums for the operational phase of the infrastructure project. Furthermore, the consequences of not adhering to the express obligations were also identified. This standardised contract approach to PF2 provided for a more robust transaction, allowed stakeholders to be clear about their contractual obligations and identified the risks specific to the project.

The PF2 procurement model continued to adopt a successful feature of PFI, namely the consolidation of construction and asset management into one contract. However, these new contracts were much more flexible; for example, soft facilities management services (cleaning and catering) were now optional under the PF2 model. The standardised approach of PF2 and the introduction of flexible asset management arrangements provided a proactive approach to managing risks in infrastructure projects.

3 Authority as co-investor

As the public sector stakeholder engages in their new role as the 'authority', they become a co-investor in the infrastructure project. As co-investor, they have a board position in the private sector consortia that facilitates their ability to directly engage with financial institutions including institutional investors, banks and credit rating agencies. This co-investor engagement permitted the public sector authority to ascertain alternative financing sources. This sharing of financial risks also served to encourage and support private sector engagement through direct investment from the authority, their unique access to capital, and their knowledge and understanding of financial investment sources and support financing investment sources.

The greater collective sharing of infrastructure project risks and the express identification of both the inherent risks and stakeholder obligations had the potential to remove private sector barriers to infrastructure investment. Additionally, the streamlined and standardised framework changes aimed to attract private sector investors by removing unnecessary procurement and contract bureaucracy. From a public sector stakeholder perspective, the PF2 framework changes further encouraged consideration of PF2 as an infrastructure procurement framework.

Ultimately, PF2 sought to enhance value for money derived from infrastructure projects through a variety of risk reduction and risk management measures, thus creating a sustainable procurement model for infrastructure investment.

Challenges to PF2

As discussed, a key feature of PF2 includes a life cycle gain-share mechanism, designed to reduce risk and improve value for money. To maintain off-balance sheet accounting for PF2 projects, the HM Treasury subsequently changed the gain-share mechanism from 50 per cent to 33 per cent (NAO Report, 2018). These challenges resulted in a very subdued uptake of PF2. Since the launch of PF2 in 2012, only six PF2 projects had reached financial close. It was originally anticipated that PF2 would raise £1.75bn of financing, but this was later reduced to £623m. UK banks were also hesitant to provide long-term infrastructure financing due to onerous new Basel III requirements. Ergo, there was limited interest in procuring infrastructure using the PF2 model.

In November 2018, the Chancellor of Exchequer announced that the UK would no longer use PFI or PF2 as a model of infrastructure procurement. This decision was made following the high-profile collapse of Carillion. Carillion were appointed under a PFI contract to develop and run several public services for both the NHS and Ministry of Defence. Carillion collapsed in January 2018 with an estimated debt of £1.5bn. Existing PFI and PF2 projects will continue, however.

The then Chancellor, Philip Hammonds, concluded his autumn 2018 statement by stating that:

> half of the UK's £600bn infrastructure pipeline will be built and financed by the private sector. The government will continue to develop new revenue support models and consider how existing models – such as the Regulated Asset Base model and Contracts for Difference – can be applied in new areas, and remains open to new ideas from the market. The government will not reintroduce the private finance initiative model (PFI/PF2).

There are three key reasons for using PPPs to deliver infrastructure according to Kwak et al. (2009): 1. The reduction of the financial burden on public bodies; 2. The transition of risk away from the public sector; and 3. The increase in efficiency, value for money and reliability of the services and assets provided. However, Kwak et al. (2009), Noring (2019), and Warsen et al. (2018) all note that many PPPs fail to deliver the conceptual benefits predicted prior to entering into the partnership. There are many reasons for this, including: 1. Differing expectations of the public and private sector actors; 2. Poor Government objectives and ongoing commitment (fluctuations in buy-in from those within elected positions); 3. The complexity of the decision-making processes; 4. Inadequate frameworks and policies; 5. Inadequate risk management; and 6. Poor transparency Ultimately the difficulty for PPPs has always been (and possibly always will be) the balancing of the core purposes of both public and private participants. The public sector has to act in the best interests of the community it serves, while the private sector has the goal of maximising returns on invested capital in line with shareholder expectations.

Priority School Building Programme (PSBP)

The first confirmed programme to which PF2 was applied was the £1.75bn privately financed element of the PSBP, which replaced the Building School Framework (BSF) infrastructure procurement model. The UK Government announced that the PSBP model would deliver £4.4bn in rebuilding and refurbishing school buildings in the worst condition across the country. Specifically, under phase 1 of the PSBP programme, 46 of the 260 schools identified were to be rebuilt/refurbished using private finance.

To deliver the 46 schools, local government partnered with private investors for finance, thereby adopting a PPP framework. The framework grouped schools into batches of five to boost cheaper finance arrangements. In May 2014, the Government announced a further £2bn for a second phase, PSBP2, which undertook rebuilding and refurbishment projects across 277 schools from 2015 to 2021. The Government believes that this framework delivers schools faster and cheaper.

Since the introduction of the PSBP, the UK Government has invested £11.3 billion to maintain and improve the condition of school estates. In acknowledging the success of the programme, a new 10-year programme of school rebuilding was announced by the Government in 2020. This new investment programme forms part of the Government's ambitious build back better agenda to deliver a high-quality education. Furthermore, the new build school projects will be designed to deliver on Government's climate target to net zero (Department for Education, 2021).

Case Study – Reigate Park Primary School

Reigate Park Primary School is the first of six schools to reopen in Derby through the £4.4 billion Priority School Building Programme (PSBP).

The project at a glance:

Dates: Opened September 2015

Cost: £4.6m

Project goals:

- 2,419 m^2 GIA
- Two-storey, 420 place Primary School
- 26 place Nursery

Project clients:

Derby local council, Education Authority and the selected contractor Bowmer & Kirkland.

Project benefits:

The benefit of the Priority School Building Programme has the potential to transform the learning environment for tens of thousands of pupils and their teachers while delivering value for money for the taxpayers.

NHS Local Improvement Finance Trust (LIFT) and Express LIFT (eLift)

The NHS LIFT initiative was introduced in England in 2001 (NAO, 2005) to leverage private sector investment in regional primary care projects. The LIFT model created a long-term joint venture between the public and private sectors. In 2008, the Express LIFT (eLIFT) model was introduced to replace LIFT. Essentially, the eLIFT model simplifies the contract award process by selecting a private sector joint venture partner and adopting the framework approach to infrastructure procurement.

In January 2018, Chris Whitehouse (Chairman of the LIFT council) stated that LIFT 'delivered more than 330 fit-for-purpose buildings, and generated more than £2.5bn of investment in the primary care estate' (Toogood, 2018). The LIFT model was previously endorsed in Sir Robert Naylor's review of NHS estates (2017), which considered the LIFT model to be an appropriate investment model, adding '£10bn in capital requirements [has been identified] across the Sustainability and Transformation Plans, and calls for healthcare leaders to take advance of private sector investment (under models such as LIFT) in the continuing environment of public capital restraint'. However, LIFT is not without risks. Naylor's review (2017) on NHS Property and Estates warned that the success of LIFT 'requires a robust and realistic timeframe that integrates local strategic investment plans'. Thus, clear NHS project pipelines can in turn boost private investor confidence in the LIFT market.

Devolved infrastructure procurement frameworks

This section explores the infrastructure procurement models adopted by the devolved Governments in Scotland, Wales, and Northern Ireland. The UK is a devolved nation and each devolved region has responsibility for setting PPP policy. To assist with PPP implementation across the UK, HM Treasury publishes key policy, guidance, advice and statistics on PPP to ensure the delivery of value for money, sustain market confidence and deliver improved operational performance of projects. In 2020 the HM Treasury published the *National Infrastructure Strategy* called *Fairer, Faster, Greener* in which it detailed regionally significant infrastructure schemes (refer to Figure 1.10 earlier in this volume).

Furthermore, the National Infrastructure and Construction Procurement Pipeline identifies the following regional infrastructure projects as shown in Figure 4.2.

The publication of regionally significant infrastructure schemes is aimed at 'levelling up' infrastructure across the UK and in addition to the regional schemes the publication also highlights the infrastructure projects that unite the regions. The publication specifically mentions:

- Leaving no business or community behind: A new £4 billion cross-departmental Levelling Up Fund that will invest in local infrastructure in England (which will attract funding for Scotland, Wales and Northern Ireland in the usual way);
- Creating regional powerhouses, making cities the engines of growth and revitalising towns;
- Connecting the regions and nations of the UK, and creating a united and global Britain;
- And changing how decisions are taken: Increasing the UK government's ability to invest directly in Scotland, Wales and Northern Ireland through the UK Internal Market Bill.

Source: HM Treasury (2020)

The procurement pipeline contains projects and programmes distributed across the UK but the majority of the value of the procurement pipeline relates to spending in England. This is because most infrastructure spending in Scotland, Wales and Northern Ireland is the responsibility of each devolved administration. For the purpose of this publication, where applicable, the IPA has included procurements set out in:

- The Northern Ireland Executive's Investment Strategy for Northern Ireland 2011-2021 sets out the forward programme for investment in public infrastructure. Details of government funded infrastructure contracts that have not yet entered procurement are also updated quarterly in the Infrastructure Investment Pipeline.[8]
- The Scottish government published an Infrastructure Investment Plan with a Project Pipeline in 2015 with an updated Project Pipeline in April 2020.[9]
- The Welsh government published an Infrastructure Investment Plan in 2012 with an updated Project Pipeline published in November 2019.[10]

Sector	Devolved administration		
	Scotland	Northern Ireland	Wales
Road	Devolved responsibility	Devolved responsibility	Devolved responsibility
Rail	The Scottish government is responsible for internal services. The UK government is responsible for cross-border daytime services	Devolved responsibility	Not devolved
Airports	Devolved responsibility. The regulation of air services is a reserved matter	Devolved responsibility	Devolved responsibility,
Ports	Devolved responsibility, with some minor exceptions	Devolved responsibility	Devolved responsibility, with some minor exceptions
Energy	Not devolved	Not devolved	Not devolved
Communications	Not devolved	Not devolved	Not devolved
Water	Devolved responsibility	Devolved responsibility	Devolved responsibility
Flood Defence	Devolved responsibility	Devolved responsibility	Devolved responsibility
Waste	Devolved responsibility	Devolved responsibility	Devolved responsibility
Housing	Devolved responsibility	Devolved responsibility	Devolved responsibility

Figure 4.2 Devolved procurement pipeline
Source: HM Treasury (2020).

Finally, the publication identifies the challenges and opportunities that have presented the UK during the Covid-19 pandemic. The Government has responded by determining that investment in infrastructure will 'have a key role to play in the recovery, both by maintaining jobs in the short term, and creating the conditions for long-term sustainable growth.' Testament to this commitment, in November 2020 the Government accelerated £8.6 billion of capital investment in infrastructure projects (HM Treasury, 2020).

Devolved infrastructure procurement models are outlined below.

Scotland

This section examines the infrastructure procurement frameworks adopted by the Scottish Government, which are managed and delivered by the Scottish Government and the Scottish Futures Trust (SFT).

Scottish Futures Trust (SFT)

The SFT is an infrastructure delivery company owned by the Scottish Government. It is a joint venture between public sector clients and the private sector for the delivery of regional infrastructure projects. The SFT (2016) identifies their core agenda as:

1 Plan future infrastructure investment;
2 Innovate to secure new ways to fund essential infrastructure;
3 Improve the management of existing properties; and
4 Deliver important infrastructure projects.

To bring about its core agenda, the SFT utilises a range of innovative public and private procurement models to deliver their infrastructure investment needs. According to the SFT (2021, October), these innovative infrastructure procurements have delivered a number of key benefits: the value of public infrastructure projects has exceeded £250; private capital investment has continued to be secured for more than £2.4bn worth of projects on site; and there has been a reduction of infrastructure-related CO2 emissions by 12,000 tonnes – working towards a net-zero carbon economy. This investment has supported economic growth across this devolved region.

SFT hubCos

The SFT works with local hub private sector partners in five geographical areas in Scotland to improve the delivery and management of infrastructure projects. In each of these five areas, a joint venture between the Scottish region and the private sector consortium is formed and collectively known as a 'hubCo'. The SFT have created five hubCos across Scotland. Each of the hubCos takes a long-term planning approach to identify the buildings it needs to support the delivery of improved community services. According to SFT (2021), 'hubCos are developing and delivering a diverse pipeline of best-value, award-winning community infrastructure, currently valued at more than £3bn which is anticipated to grow to over £3bn in the coming years'.

SFT procurement frameworks

The SFT has developed the following innovative infrastructure procurement frameworks:

- Non-Profit Distributing (NPD) model;
- Tax Incremental Financing (TIF);
- National Housing Trust (NHT);
- Growth Accelerator (GA); and
- Mutual Investment Model.

NON-PROFIT DISTRIBUTING (NPD) MODEL AND RISKS

The NPD model superseded the traditional PFI model in Scotland. The NPD model is an innovative delivery consortia procurement framework that delivers investment in three key infrastructure sectors: further education, transport and health.

A SPV consortium is created that facilitates shares in a limited company. According to the SFT (2016), the shares in the project company are held by the private sector investors except for one 'golden share' held by the authority, i.e. the role of the public sector is enhanced to that of a co-investor, which facilitates greater transparency and accountability for the infrastructure projects and reduces the risks associated with the delivery of poor value for money.

One of the inherent risks of PFI was the potential for windfall profits that could be generated for and by the private sector, and thus the erosion of value for money. The NPD framework complies with current procurement rules by adopting an open procurement process; however, this process seeks to eliminate uncapped equity returns for private investors, which became an issue with PF2.

The NPD framework redirects excess profits through reinvestment into public services. This significant feature of the NPD model has reduced risks associated with excess profit by ensuring maximum value for money is derived from the infrastructure project.

In keeping with a successful features of PF2, the NPD adopts a 'simplified standard contract', which improves the costs and time in procuring the partners. Excessive procurement time and costs are considered to be a significant barrier to private investment, which the standardised contract has mitigated. According to SFT (2015), the NPD procurement model contains the following risk reduction features which were absent in its failed predecessor, the PFI:

- Optimum risk allocation.
- Whole-life costing.
- Maximised design efficiencies.
- Robust programming of life cycle maintenance and facilities management.
- Performance-based payments to the private sector.
- Single point delivery system, which creates a more effective means of communication and can minimise the risk of miscommunication, reducing interface risk for the public sector client. It is evident that the public interest is represented in the governance of the NPD structure, which increases transparency and accountability, and facilitates a more proactive and stable partnership between public and private sector parties.
- Improved service provision.
- Capped returns ensure that a 'normal' level of investment return is made by the private sector and that these returns are transparent. Capping profit returns aims to minimise the risk of excessive private sector profits and ensure public sector accountability, while the transparency of profits seeks to minimise the risk associated with the lack of competitive investor appetite, which

was in part due to private sector concerns regarding the uncertainty of profits expected to be generated from investing in public infrastructure projects. Thus, transparency in investment returns seeks to remove private investor uncertainties regarding profit.

- Operational surpluses generated by the project company are directed to the public sector client or a third party nominated by it. A key feature of NPD is the reinvestment of excess profits in public services, enhancing value for money on these infrastructure projects.

According to the Audit Scotland (2020), the NPD framework and hub private financing have supported £3.3 billion of additional investment in public assets with another £5.7 billion of investment also being supported under earlier PFI contracts. Private finance costs more than traditional forms of financing, yet the Scottish Government has accepted these costs to enable additional investment to take place. It is not really clear, however, 'how decisions have been taken about which projects will use private finance, or how well this is achieving the best balance of cost and benefits in practice'. See Figure 4.3 for a summary of infrastructure investment models in Scotland.

Changes to the interpretation of national accounting rules mean the Scottish Government has now stopped using the NPD and hub forms of private financing for new projects. NPD's use of private sector borrowing to build infrastructure projects is more expensive than using public sector sources of borrowing. And, despite the challenge in securing private investment whilst maintaining public accountability, after the SFT appraisal of infrastructure procurement models in 2019, the Scottish Government has decided to adopt the Mutual Investment

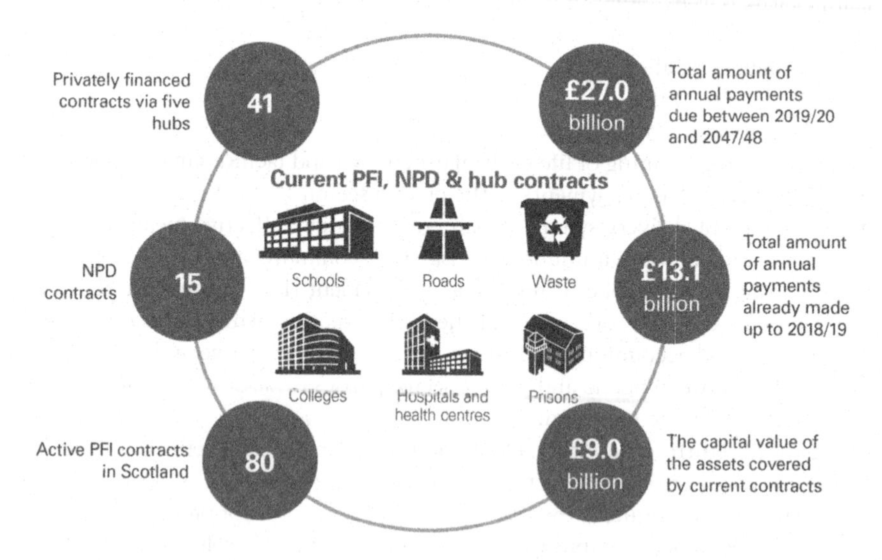

Figure 4.3 Infrastructure investment models in Scotland: Summary
Source: Audit Scotland (2020).

Model (MIM) – known as the Welsh model (refer to Welsh devolved administration section of this chapter for further details on the MIM) – for infrastructure works.

Audit Scotland (2020) also cited guidance on the adoption of privately financed infrastructure:

> - better document and report how decisions on the use of private finance are made at a programme level, and how the overall combination of programme and project funding aims to maximise investment and benefits;
> - better communicate the rationale of project financing and funding decisions to public sector organisations and Parliament;
> - continue to monitor existing NPD and hub projects to review if the models are successfully achieving their original aims, and documenting lessons learned;
> - set out how the MIM will operate, and establish clear criteria for selecting programmes and monitoring risks;
> - develop its public reporting to provide more information on the costs and benefits of using private finance, the management of risks and outcomes delivered, and its contribution to supporting economic policies and growth.
>
> Audit Scotland (2020)

At the time of writing the MIM is still the procurement model adopted by Scotland.

TAX INCREMENTAL FINANCING (TIF)

TIF is also a procurement innovation that relies on private finance to fund local infrastructure. Unlike other PPP models, however, investors do not finance projects directly but fund local infrastructure projects through business revenues such as rates. Fundamental to the success of a TIF project is the requirement that the private investor must demonstrate a contribution to economic growth and regeneration. It is hoped that TIF will realise further private sector investment.

According to SFT (n.d.), there are currently four TIF projects:

- Glasgow City Council's City Centre TIF has the capacity to deliver significant benefits to the City Centre;
- Fife Council will invest TIF enabling funding to improve vehicle and marine access to Energy Park, Fife, allowing the potential for further renewable, offshore and wind project capability to be developed;
- Argyll & Bute Council's Lorn Arc TIF is aimed at further enabling Oban's wider economy and ensuring the economic infrastructure which will act as a catalyst to grow marine tourism, renewables and aquaculture is in place;
- Falkirk Council is investing to support commercial activity in manufacturing, distribution and support sectors in Grangemouth. This will include strategic road improvements and flood defences to enhance and protect the region.

HOUSING AND DEVELOPMENT MODELS

The SFT have developed niche procurement models for social housing infrastructure. These models focus on all streams of the housing market and deliver homes across Scotland with an emphasis on affordable housing. They are applicable to:

- Home Ownership Made Easy (HOME)
- Enabling strategic housing sites
- Housing in town centres

In order that Scotland could finance the demand for new homes, the SFT developed a range of new housing solutions aimed at delivering new and sustainable housing demand until 2040. These solutions have evolved from the regional National Housing Trust (to date it has delivered 1,731 homes) and housing delivery partnerships to a 'Home Ownership Made Easy' (SFT, August 2020). According to SFT, Home Ownership Made Easy (HOME) 'is a new way of delivering affordable homes that provides an easier route to home ownership for first-time buyers and also offers an alternative option for older homeowners who want to move to a house that is right for their needs' (SFT, n.d.).

GROWTH ACCELERATOR (GA)

The Growth Accelerator (GA) initiative is a city centre framework designed to enhance and maximise investment and growth in cities. The SFT (2021) documented that in the City of Edinburgh, the GA framework has realised an investment of £60m in public infrastructure, which is seeing £850m of private sector investment being made into the heart of the city.

Wales

Mutual Investment Model (MIM)

Following on from the success in Scotland of the role out of PPP models, in 2017 the Welsh Government introduced their version of PPP called the Mutual Investment Model (MIM). Whilst acknowledging the constraints of the NPD Scottish model, MIM is not considered as capital expenditure. As in Scotland, the Welsh devolved Government required a significant investment in infrastructure and realised the benefits of collaboration with the private sector. Announcing the MIM in the Welsh Assembly, Cabinet Secretary Mark Drakeford (Ashurst, 2017) 'made a commitment to invest approximately £1 billion in capital infrastructure investment through MIM, in the transport, health and education sectors'. Whilst MIM is a PPP model, it differs from PFI, PF2 and NPD in that the private investor is also required to create apprenticeships and traineeships, which benefit the community (Welsh Government, 2017).

According to the Welsh Government (2017), the key features of MIM mitigate the inherent weakness of all previous PPP models (lack of transparency, failure to deliver value for money, time and cost overruns and the lengthy procurement processes), adding that MIM provides:

1. The opportunity for the Welsh government, as the 'authority', to take a minority equity stake (up to a maximum of 20% of issued share capital) in Hold Co. (consisting of key stakeholders), which facilitates board-level access and creates a transparent framework. Crucially, it also facilitates greater risk sharing between the private sector and the authority. This may attract greater competition from potential private sector investors. The creation of Hold Co. also can facilitate 'third-party equity' funding competitions for a specified portion of equity in Hold Co. at the preferred bidder stage.
2. Enhanced stakeholder involvement through the Welsh government's ability to appoint a director to the boards of Project Co. (private sector) and Hold Co., and the option for them to also appoint an observer to those boards. The authority as stakeholder aids transparency throughout all phases of the infrastructure project and enhances the accountability of Project Co. and Hold Co. Greater accountability and transparency in the process help to identify risks and challenges, and to collectively mitigate risks as they occur. Robust scrutiny of the infrastructure project during both construction and operations is undertaken by Project Co. (including the authority with its Project Co. board position). This feature, combined with stringent project performance measuring and monitoring, seeks to ensure that the public interest is served and protected.
3. No 'soft' services on accommodation projects. Like PF2, the MIM framework includes a more flexible approach to the asset management of infrastructure projects. Realising the risks and failure of soft service contracts during the asset management phase of the project, the MIM has moved these services from the private sector to the public sector, particularly on accommodation projects.
4. The ability of the authority to carry out low-value changes and the adoption of a rigorous change procedure. The increased flexibility introduced by the PF2 model has been included in the MIM framework. Previously, financial risks incurred by the authority when making changes during the construction and management of the infrastructure project created excessive cost penalties in procurement. The introduction of a change procedure should reduce the fiscal impact of changes.
5. A key characteristic of the MIM framework is the mandatory inclusion of wider local economy benefits, identified as 'community benefits'. Community benefits under the MIM framework include the creation of long-term apprenticeships and traineeships beyond the construction phase of the infrastructure project.

MIM SCHEMES

The Welsh Government (2018) set out to invest over £6.5bn in capital funding over the period 2018–19 to 2020–21, which was further boosted through finance initiatives, borrowing and other sources of funding. In the spring of 2017, the Cabinet Secretary for Finance and Local Government announced the Government's intention to undertake three schemes using the MIM framework:

- Redevelopment of the Velindre Cancer Centre, Cardiff; onstruction to commence in 2023;
- Work to widen the A465 from Dowlais Top to Hirwaun to two lanes; commenced in 2021 and due to complete in 2025; and
- The 21st Century Schools and Colleges Programme, which will begin in 2022 and includes funding through traditional capital and £500m in investment enabled through revenue funding via the MIM.

Northern Ireland

The Northern Ireland Assembly is responsible for legislation pertaining to matters in Northern Ireland, as well as advising and scrutinising local government departments. In 2003, it established the Strategic Investment Board (SIB) Northern Ireland. The role of the SIB is to prepare the Investment Strategy for Northern Ireland (ISNI) on behalf of the NI Executive. The ISNI identifies major projects and infrastructure programmes across Northern Ireland, which the SIB plans and delivers in conjunction with Government departments. Currently, the SIB supports more than 50 projects and programmes across a wide range of business areas and

> since the development of the last ISNI in 2011, over £14.9 billion has been spent in maintaining, upgrading and extending our regional infrastructure, an average of £1.48 billion per year, or £8,000 per person. Over 400 projects have been completed, thirty with a value of more than £100 million.
>
> SIB (2020)

According to the SIB (2011), most of the finance for these projects came from the NI Block Grant. However, there are 39 operational PFI/PPP schemes in Northern Ireland; these have delivered capital assets financed from private sources. Of these 39 schemes, some have concluded and others still have more years to run on their contract. After extensive criticism of the PFI scheme and its subsequent withdrawal, however, there has been no new infrastructure procurement models introduced in Northern Ireland despite the UK-wide introduction of PF2 and later successful initiatives.

On 26 January 2017, the Northern Ireland Assembly was suspended and despite fresh elections being held on 2 March 2017, negotiations between the two main parties failed to reach agreement. With funding allocations in the hands of the Northern Ireland civil service, the UK Government passed a budget for Northern Ireland for the 2017–18 fiscal year. On the 11 January 2020 the Northern Ireland

Assembly returned, three years after it was suspended. Almost immediately Northern Ireland found itself in the throes of the Covid-19 pandemic. Hence, much of the work of the NI executive has been in dealing with the health, social and economic impacts of the pandemic.

The Executive Office in Northern Ireland determines and prioritises infrastructure investment. These decisions are based on the Investment Strategy for NI produced by the Strategic Investment Board and in collaboration with all regional government departments and regional Strategic Investment Board (SIB). Furthermore, it is their role to consider the value of PPP in delivering regional infrastructure.

Belfast Local Development Plan 2035

Like Scotland, NI also seeks to enhance the economic viability and growth in regional hubs and cities. The premier city in NI, Belfast, has secured a City Deal investment which seeks to deliver on the 'Belfast Agenda', which is an aspirational vision of Belfast based on Belfast City Council's Local Development Plan 2035. According to Belfast City Council (2017), central to the Belfast Local Development Plan 'is to make the Belfast region a global investment destination' and 'create 20,000 new and better jobs in the sectors where we are experiencing most rapid growth and have potential to become world leaders'. It will focus on four key investment pillars: digital and innovation, infrastructure, tourism-led regeneration, and skills and employability. The Belfast Region City Deal provides seed funding to promote and accelerate the Plan and was signed at the end of 2021.

Summary

This chapter presented a timeline of PPP from its introduction in the mid-1990s to its radical redesign of the last few years. The key factors that led to the radical changes in PPP models of procurement are multifaceted, complex and difficult to resolve. However, the case remains for PPP as whether it is a necessary infrastructure investment model. The challenge has always been to ensure that the PPP delivers value for money. Significant recent changes have enhanced and transformed the PPP models of procurement, from the now redundant PFI and PF2 models, to the new show in town – the Mutual Investment Model (MIM). Significant too is the roll out of PPP models by the devolved regions. Both nationally and regionally the UK Government recognises the benefits of a PPP in the delivery of essential infrastructure on which growth and prosperity depend.

References

Anglian Water (2018). *Our Plan 2020–2025*. PR19. Available at: https://www.anglianwater.co.uk/siteassets/household/about-us/01-pr19-our-plan-2020-2025.pdf

Ashurst (2017). "A new model for Welsh infrastructure: 'MIM's' the word!". *Infraread*, issue 10, 3 October. Available at: https://www.ashurst.com/en/news-and-insights/insights/mutual-investment-model/

Audit Scotland (2020, January). *Privately Financed Infrastructure Investment: The Non-Profit Distributing (NPD) and Hub Models*. Available at: https://www.audit-scotland.gov.uk/uploads/docs/report/2020/nr_200128_npd_hubs.pdf

Belfast City Council. (2017). *The Belfast Agenda: Your Future City*. Belfast's Community Plan. Belfast: Belfast City Council. Available at: https://www.belfastcity.gov.uk/belfastagenda

Budget 2018: Philip Hammond's speech. Available at https://www.gov.uk/government/speeches/budget-2018-philip-hammonds-speech

Cabinet Office (2020, November). "Procurement Policy Note – Introduction of Find a Tender". Available at: https://assets.publishing.service.gov.uk/government/uploads/system/uploads/attachment_data/file/937209/PPN_08_20_Procurement_Policy_Note_Introduction_of_Find_a_Tender.pdf

Department for Education (2021, July). *Prioritising Schools for the School Rebuilding Programme*. Government Consultation. Available at: https://consult.education.gov.uk/school-rebuilding-division/prioritising-schools-for-the-school-rebuilding-pro/supporting_documents/School%20Rebuilding%20Programme%20Consultation.pdf

Department of Health and Social Care (2017). "The Naylor Review". Available at: https://www.gov.uk/government/publications/nhs-property-and-estates-naylor-review

Education and Skills Funding Agency (2014–). "Priority School Building Programme". Collection. Available at: https://www.gov.uk/government/collections/priority-school-building-programme-psbp

EU Directive 2004/18. Available at: https://www.legislation.gov.uk/eudr/2004/18/pdfs/eudr_20040018_adopted_en.pdf and https://www.legislation.gov.uk/eudr/2004/18/contents

Garvin, M. J. and Bosso, D. (2008). "Assessing the effectiveness of infrastructure public–private partnership programs and projects", *Public Works Management & Policy*, vol. 13, pp. 162–178.

Hellowell, M., Price, D. and Pollock, A.M. (2008). "The use of Private Finance Initiative (PFI) Public Private Partnerships (PPPs) in Northern Ireland". Northern Ireland Public Service Alliance. Available at: https://allysonpollock.com/wp-content/uploads/2013/04/NIPSA_2009_Hellowell_PFIPPPNIreland.pdf

HM Treasury (2012, December). *A New Approach to Public Private Partnerships*. Available at: https://assets.publishing.service.gov.uk/government/uploads/system/uploads/attachment_data/file/205112/pf2_infrastructure_new_approach_to_public_private_parnerships_051212.pdf; followed by *A New Approach to Public Private Partnerships: Consultation on the Terms of Public Sector Equity Participation in PF2 Projects* (2013, July). Available at: https://assets.publishing.service.gov.uk/government/uploads/system/uploads/attachment_data/file/207382/pf2_userguide.pdf

HM Treasury (2012, December). *PF2: A User Guide*. Available at: https://assets.publishing.service.gov.uk/government/uploads/system/uploads/attachment_data/file/207382/pf2_userguide.pdf

HM Treasury (2013, January). *Infrastructure Procurement Route Map: A Guide to Improving Delivery Capability*. Available at: https://assets.publishing.service.gov.uk/government/uploads/system/uploads/attachment_data/file/329052/iuk_procurement_routemap_guide_to_improving_delivery_capability_280113.pdf

HM Treasury (2015). "Priority Schools Building Programme", HM Treasury, London.

HM Treasury (2019). *Private Finance Initiative and Private Finance 2 Projects*. Available at: https://assets.publishing.service.gov.uk/government/uploads/system/uploads/attachment_data/file/805117/PFI_and_PF2_FINAL_PDF1.pdf

HM Treasury (2020, November). *National Infrastructure Strategy: Fairer, Faster, Greener*, CP 329. Available at: https://assets.publishing.service.gov.uk/government/uploads/system/uploads/attachment_data/file/938049/NIS_final_web_single_page.pdf

HM Treasury & Infrastructure UK (2013, January). *Infrastructure Procurement Route map: A Guide to Improving Delivery Capability*. Available at: https://www.pppforum.com/sites/default/files/iuk_procurement_routemap_guide_to_improving_delivery_capability_280113.pdf

Hodge, G. and Greve, C. (2007). "Public–private partnerships: An international review", *Public Administration Review*, vol. 67, pp. 545–558.

Infrastructure and Projects Authority (2021, August). *Analysis of the National Infrastructure and Construction Pipeline*. Available at: https://assets.publishing.service.gov.uk/government/uploads/system/uploads/attachment_data/file/1016759/Analysis_of_the_National_Infrastructure_and_Construction_Pipeline_2021.pdf

Kwak, Y. H., Chic, Y. Y. and Ibbs, C. W. (2009). "Towards a comprehensive understanding of public private partnership for infrastructure development", *California Management Review*, vol. 52, no. 2, pp. 51–78.

Local Government Association (2018, January). *National Construction Category Strategy for Local Government Effective Construction Frameworks*. Available at: https://www.local.gov.uk/sites/default/files/documents/Construction%20Category%20Strategy%20Final.pdf

Local Government Association (2020). *National Construction Category Strategy for Local Government*. Available at: https://constructingexcellence.org.uk/wp-content/uploads/2020/12/National-Construction-Strategy_2020-Edition.pdf

NAO (2005, May). "Department of Health – innovation in the NHS: Local Improvement Finance Trusts". National Audit Office. 19 May. Available at: https://www.nao.org.uk/report/department-of-health-innovation-in-the-nhs-local-improvement-finance-trusts/

Noring, L. (2019). "Public asset corporation: A new vehicle for urban regeneration and infrastructure finance", *Cities*, vol. 88, pp. 125–135.

Public Sector Newsletter (2015, September). "Derby school reopens in multi-million pound hi-tech building". 21 September. Available at: https://www.public-sector.co.uk/article/c0c73d1b45bffc5dcf8290eac83a17ce

Scottish Futures Trust (2015, March). *NPD Model: Explanatory Note*. Available at: https://www.scottishfuturestrust.org.uk/storage/uploads/Explanatory_Note_on_the_NPD_Model_(Updated_March_2015).pdf

Scottish Futures Trust (2016). See: https://www.scottishfuturestrust.org.uk/

Scottish Futures Trust (2020, August). *Can Local Authorities Deliver Housing for Sale or Market Rent in Scotland?* Analysis. Available at: https://www.scottishfuturestrust.org.uk/storage/uploads/canlocalauthoritiesdeliverhousingforsale.pdf

Scottish Futures Trust (2021). "Growth Accelerator". See https://www.scottishfuturestrust.org.uk/page/growth-accelerator

Scottish Futures Trust (2021, October). *SFT Outcomes: Demonstrating Progess with Impact*. Available at: https://outcomes.scottishfuturestrust.org.uk/

Scottish Futures Trust (n.d.). "Development & Housing: Home Ownership Made Easy (HOME)". See: https://www.scottishfuturestrust.org.uk/page/home-ownership-made-easy-home

Scottish Futures Trust (n.d.). "Development & housing: Tax incremental financing". Available at: https://www.scottishfuturestrust.org.uk/page/tax-incremental-financing

Scottish Futures Trust (n.d.). "Housing". See https://www.scottishfuturestrust.org.uk/page/housing

Strategic Investment Board (2011). "Investment Strategy". Available at: http://nia1.me/3jf

The Strategic Investment Board (2020). "Investment Strategy Northern Ireland". Available at: https://sibni.org/home/investment-strategy-isni/

Toogood, D. (2018, February). "Councillor tells MPs 'partnership approach needed' over future NHS properties". *Island Echo*, 1 February. Available at: https://www.islandecho.co.uk/councillor-tells-mps-partnership-approach-needed-future-nhs-properties/

Van Ham, H. and Koppenjan, J. (2001). "Building public–private partnerships: Assessing and managing risks in port development", *Public Management Review*, vol. 3, no. 4, pp. 593–616.

Warsen, R., Nederland, R., Klign, E. H., Grotenbreg, S. and Koppenjan, J. (2018). "What makes public–private partnership work? Survey research into the outcomes and the quality of cooperation in PPPs", *Public Management Review*, vol. 20, no. 8, pp. 1165–1185.

Welsh Government (2017–). *Mutual Investment Model for Infrastructure Investment*. Collection. Available at: https://gov.wales/mutual-investment-model-infrastructure-investment; specifically: https://gov.wales/sites/default/files/publications/2018-07/standard-form-project-agreement-user-guide.pdf

Welsh Government (2017). "Mutual Investment Model: Guidance". Available at: https://gov.wales/mutual-investment-model-infrastructure-investment

Welsh Government (2018). *Wales Infrastructure Plan – Mid-point Review 2018*. Available at: https://gov.wales/sites/default/files/publications/2018-07/wales-infrastructure-investment-plan-mid-point-review-2018-a.pdf

WTO (2021). *Agreement on Government Procurement*. Available at: https://www.wto.org/english/tratop_e/gproc_e/gp_gpa_e.htm

5 International infrastructure investment

Introduction

This chapter considers international infrastructure investment. It comprises an examination of developed and developing economies infrastructure investment spend and future projected infrastructure investment needs. This chapter presents the investment paradigm and the identification of international funding and procurement models utilised by developed and developing countries.

Learning outcomes

In this chapter you will be able to:

1. Understand the global infrastructure investment gap and the correlation between investment and economic growth and wellbeing in both developed and developing countries.
2. Appreciate the reasons for the global investment gap and the subsequent required sustained investment needs.
3. Be aware of the infrastructure investment models and procurement methods that are applied to developed and developing countries.

Investment in infrastructure

Regardless of a country's status as either a developed or developing nation, Rutherford (2002) defined infrastructure as 'the basic services or social capital of a country, or a part of it, which make economics and social activities possible'. Comparatively, Grimsey and Lewis (2002) deemed that infrastructure comprehensively includes public services, the economic sector as well as social contributors which influence living standards and quality of life. Therefore, the provision of infrastructure is vital to nations' wealth and wellbeing. Infrastructure can be the vector to facilitate urbanisation, generate social stability and mitigate disasters (World Bank, 2014). It is therefore understandable that infrastructure investment occupies a pivotal role in governmental national development strategies. This chapter will explore the societal benefits derived from infrastructure investment.

DOI: 10.1201/9781003171805-5

Economic impact of infrastructure investment

Numerous studies have emphasised the positive correlation between infrastructure investment and the economic growth of a region or country. The research origins of this relationship stem from the seminal work by Aschauer (1989a, 1989b, 1990), who identified a strong positive correlation between infrastructure provision and macroeconomic growth in the United States. In his work, Aschauer considered the relationship between public capital expenditure and private-sector productivity. Munnell (1992) reiterated Aschauer's prior findings; infrastructure investment increases economic growth and productivity. To do this, Munnell produced a summary of findings illustrating positive and statistically significant parallels between infrastructure investment and economic growth. In 2004, De la Fuente and Estache (2004) evaluated 102 countries over a period of 15 years. Their research confirmed a positive correlation between the impacts of infrastructure on growth.

Almost two decades on, there is widespread acceptance that infrastructure correlates to growth. Recently there have been reports and publications that have quantified the growth derived from infrastructure investment. Consistently the quantitative measures adopted are based on measuring the increase in GDP. Although there persists consternation over the accuracy of using GDP, in the main, the effect did not rise above 2.5 but tended to be greater in a recession. In this context infrastructure can be said to influence not only economic growth but also working conditions, education provision, health levels, social connections, quality of life, well-being, the environment and civic engagement and governance (Catalano and Sartori, 2013).

The global infrastructure investment challenge

Globally, every nation is threatened by sustained underinvestment in infrastructure, the need to invest more to mitigate the damage of the climate emergency and to provide more infrastructure to meet the needs of growing populations. McKinsey Global Institute (2013) estimated that, on average, between 1992 and 2011 3.8 per cent of global GPD was spent on infrastructure;equating to an annual average of US$2.4trillion (about $7, 400 per person in the US). Over this timeframe, PwC (2014) estimated that there has been a change in thinking in developed and developing markets. Spending in advanced economies markedly contracted from around 3.6 per cent of GDP in the 1980s to just 2.8 per cent in 2008. Spending less means that not only is money needed to invest in new infrastructure, but it is also needed to upgrade existing infrastructure. Comparatively, emerging markets, premised on economic growth, have calibrated their capital investment strategies toward economic infrastructure development. Investment outlay on infrastructure in developing regions outlay has shifted from around 3.5 per cent of GDP to 5.7 per cent over the same period, mostly driven by the Chinese economy (PwC, 2014).

There is clarity around the need for greater and sustained infrastructure investment. Numerous studies have identified the magnitude of the infrastructure

Figure 5.1 Total forecast infrastructure investment gaps
Source: Global Infrastructure Hub (2021a).

investment gap (see Figure 5.1). PwC (2014) reviewed infrastructure spending trends up to 2025, reporting that infrastructure investment must increase from US$4tn per annum in 2012 to over US$9tn by 2025. Inderst and Stewart (2014) also appraised infrastructure requirements. Accounting for the limitations of the MGI report, they projected an increase from US$3.5tn annually to US$5tn per annum, equating to US$80tn of investment by 2030.

Global infrastructure investment

Globally, infrastructure investment can be funded both from the public sector and the private sector. With the retrenchment of infrastructure investment by central governments, there is a growing realisation that private sector investment can overcome their investment shortfall. Aspirations and pressures for better economic and social infrastructure provision are driving the expansion of infrastructure projects globally. According to Bhattacharya et al. (2016), these pressures are global population increases; the sustainable development goals (SDGs) agenda;

and the need to invest for an uncertain future, in relation to climate change. The world has responded and according to the World Bank (Fakhoury, 2022), private sector investments in infrastructure in 2021 were $76.2 billion– a 49 per cent increase from the previous year. The dominant infrastructure sector was transportation, in particular airports, with private investment in airports exceeding £20 billion in 2021. Recent trends in private infrastructure investment can be seen in Figures 5.2 and 5.3.

As can be seen in Figure 5.2, the data from GI Hub (2021b) show that global private investment in infrastructure projects in primary markets fell by 6.5 per cent in 2020; an indication that the Covid-19 pandemic has extended the infrastructure investment gap globally.

Disappointingly, investment trends in middle- and low-income countries (see Figure 5.3) show that during the last decade there has been significantly less investment than in high-income countries. The pandemic has also widened this investment gap;in middle- and low-income countries there was a further decline in private infrastructure investment which saw investment fall by 28 per cent in 2020, whereas investment in high-income countries rose by 2 per cent.

Private infrastructure investment is carried out via many different investment instruments across the various income groups. As can be seen from Figure 5.4, over the period 2012–2020, the dominant private infrastructure investment instrument used was private sector loans, in both high- and middle- and low-income countries. Yet the trend is different, with their adoption increasing in high-income countries and decreasing in middle- and low-income countries. In addition to private sector loans, other infrastructure investment instruments used by private investors include bonds, government loans, equity, grants, and more recently, green bonds.

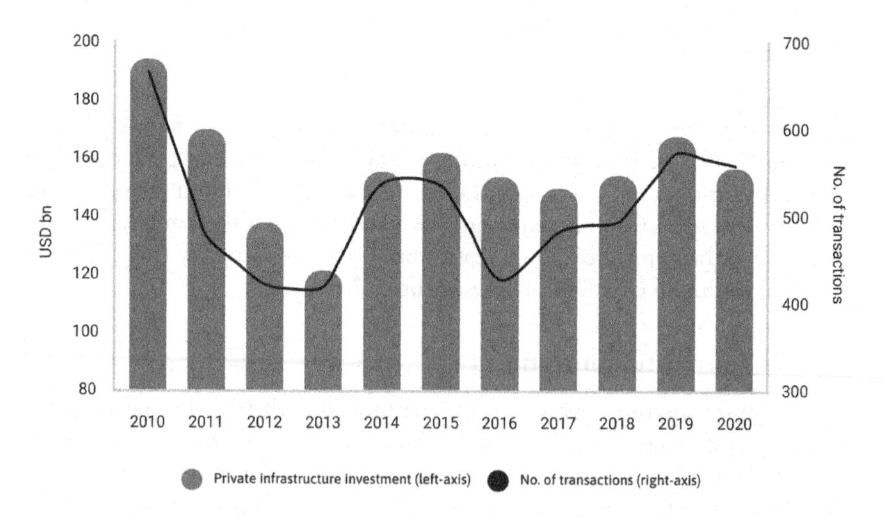

Figure 5.2 Private investment in infrastructure projects in primary markets
Source: Global Infrastructure Hub(2021b).

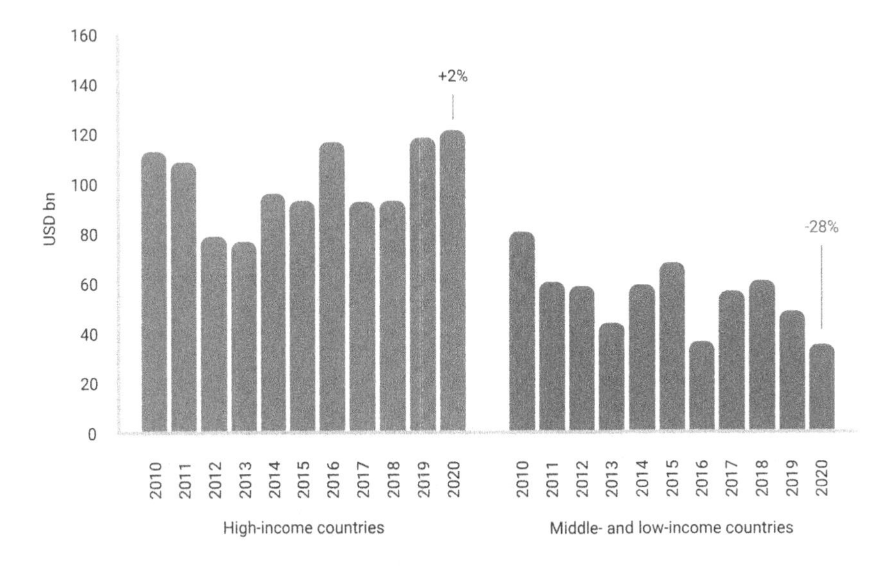

Figure 5.3 Private investment in infrastructure projects by income group 2010–2020 (US$billion and percentage growth in 2020)
Source: Global Infrastructure Hub (2021b).

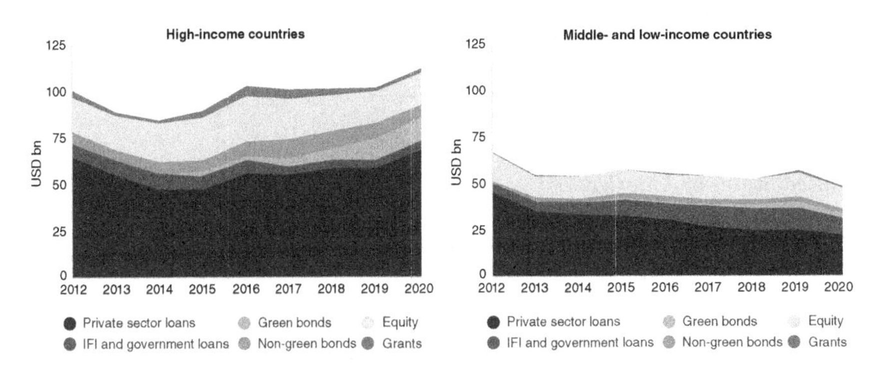

Figure 5.4 Private investment in infrastructure projects by instrument type and income group
Source: Global Infrastructure Hub (2021b).

There are many ways to finance private investment (see Figure 5.5). The majority of private infrastructure projects (63 per cent) are financed by financial services institutions, primarily commercial and investment banks. Developers are the second largest type of financier (10.2 per cent), mostly in the form of equity. Insurance companies and pension funds directly finance only 1 per cent of private investment in infrastructure projects.

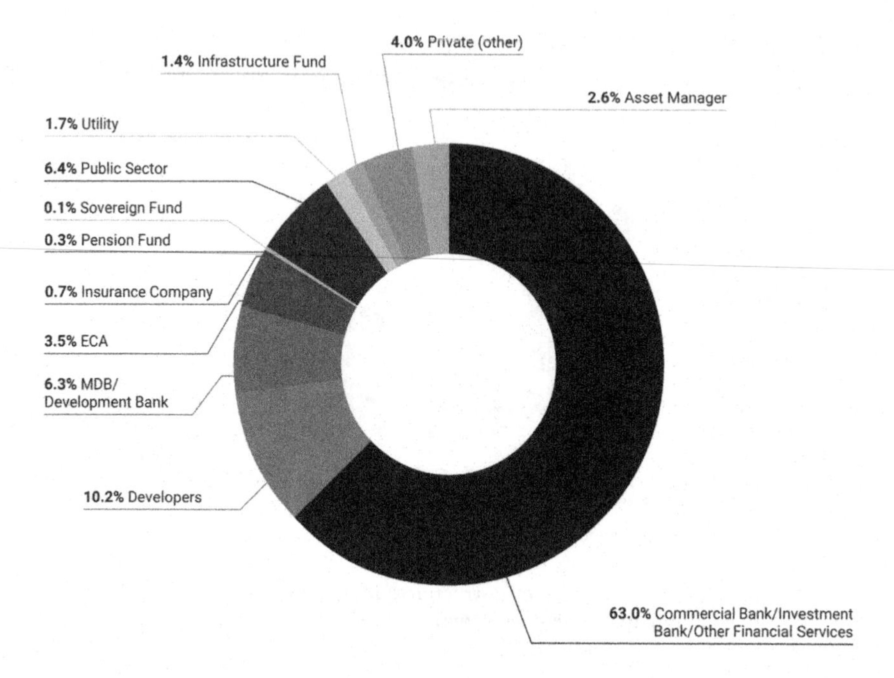

Figure 5.5 Private investment in infrastructure projects by financier, 2020
Source: Global Infrastructure Hub (2021b).

Developed economies

United States of America

According to McKinsey (2021a), growth, investment, and economic activity have been uneven across the United States. In recent years, the country's economic growth has been concentrated in just a few places. For example, only 6 per cent of US counties account for two-thirds of GDP output. However, the USA is still considered 'a dynamic and resilient economy' (McKinsey, 2021a). One approach to ensuring growth in the nation is by accelerating the infrastructure and housing agenda while simultaneously increasing its commitments to decarbonisation and net zero emissions. This collective approach signals the resetting of Government strategy of supporting green technologies and infrastructure investments.

President Biden in his first 100 days in office pledged to increase investment in infrastructure. As a result, on 15 November 2021 he signed the Infrastructure Investment Jobs Act (IIJA; also known as the Bipartisan Infrastructure Law) and pledged an approximately $1.2 trillion infrastructure investment package over the next five years (see Figure 5.6).

The IIJA was described by the Whitehouse as a 'once in a generation' investment designed to rebuild the USA's crumbling infrastructure. It will, among other things,'rebuild America's roads, bridges and rails, expand access to clean drinking water, ensure every American has access to high-speed internet, tackle the

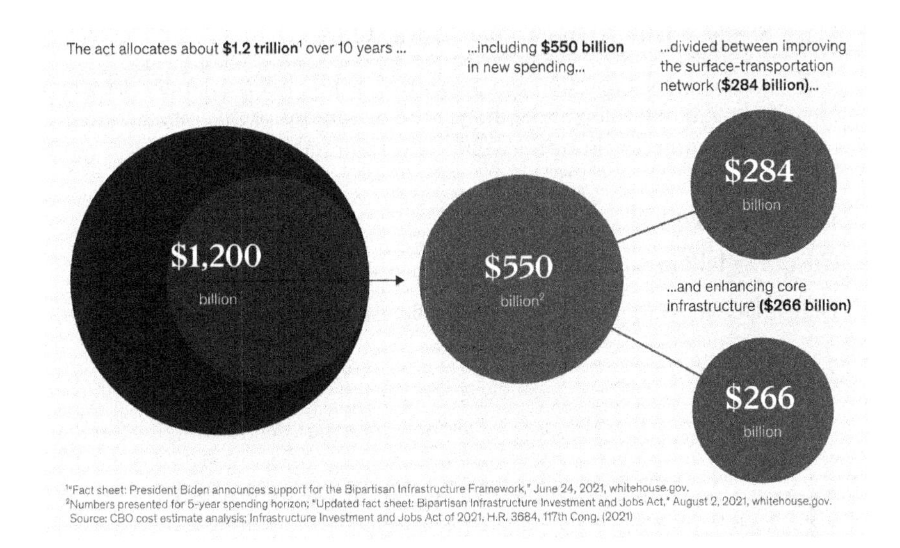

Figure 5.6 Infrastructure and Investment Jobs Act (IIJA) investments
Source: McKinsey (2021b).

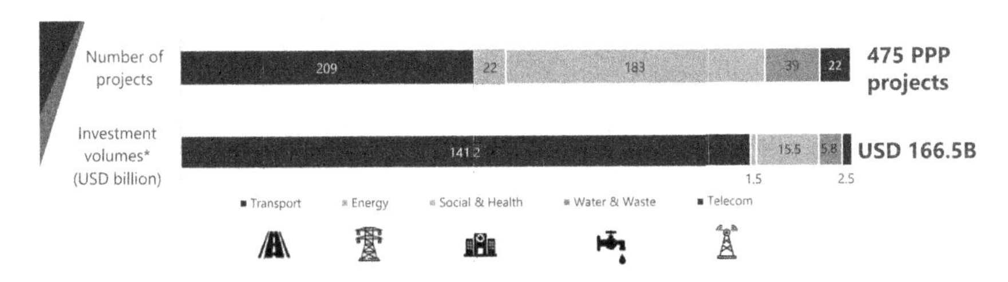

Figure 5.7 PPP project portfolio by sector (USA)
Source: InfraPPP World Project Database (2019).

climate crisis, advance environmental justice, and 'invest in communities that have too often been left behind'(Whitehouse Press Office, 2021).

In the delivery of its infrastructure programmes, the United States extolls the use of Public Private Partnerships (PPPs). Among other laws and programmes that support PPPs, recent administrations have shown their backing for private investment in infrastructure through the Fixing America's Surface Transportation Act (2015), the Investment Partnerships Program (2016) and the Project Support Fund (2017). In 2019, investment in PPP projects reached an amount of US$83.3 billion, following US$19.5 billion in 2018 and US$19.7 billion in 2017.

As can be seen in Figure 5.7, in 2018/19 transport was the dominant infrastructure investment sector (209 projects), followed by social and health (183 projects). Currently (May 2022) the USA has 773 PPP projects and a pipeline of 483 PPP projects, of which 169 projects are in planning and 314 projects are in the tender stage (InfraPPP, 2022). The passing of the IIJA will further increase the infrastructure investment pipeline.

Case Study – JFK Airport, New York, USA

The investment of $9.5 billion in a terminal at JFK International Airport, New York, will redevelop and enhance the operations of this international airport. The project is known as The New Terminal One (NTO).

Project at a glance:

Cost: $9.5 billion

Dates: Due to commence in 2022, completion expected 2030.

Project goals:

2.4 million square foot state-of-the-art new international terminal

- new arrivals and departures hall and the first set of new gates,
- 23 gates, check-in halls and arrival spaces,
- 300,000 square feet of dining and retail amenities.
- integrate modern capabilities in both sustainability and security.

Project funding:

The construction of the new terminal will be privately financed by the JFK Millennium Partners consortium. This funding agreement was reached with the Port Authority of New York and New Jersey (NTO) consortium of financial sponsors, including Carlyle, JLC Infrastructure and Ullico.

Project benefits:

Improving operations and upgrading the gates at the NTO, when complete, will create more than 10,000 jobs and transform JFK into a world-class airport worthy of New York and the region.

Throughout the development phase the NTO project seeks to support community development initiatives, local job opportunities, small business outreach and development, and educational programming for local students.

Ireland

From 2014–2019 Ireland experienced a steady and determined improvement in the economy. Public investment in the construction sector increased from €4.3 billion to €7.3 billion and private sector investment similarly increased from €8.3 billion to €18.9 billion (see Figure 5.8).

In 2019 the Department of Public Expenditure and Reform published a Project Ireland 2040 report, *Prospects – Ireland's Pipeline of Infrastructure Projects, 2019,* which considered the efficient delivery of major infrastructure projects throughout the country. Project Ireland 2040 is the Government's long-term overarching

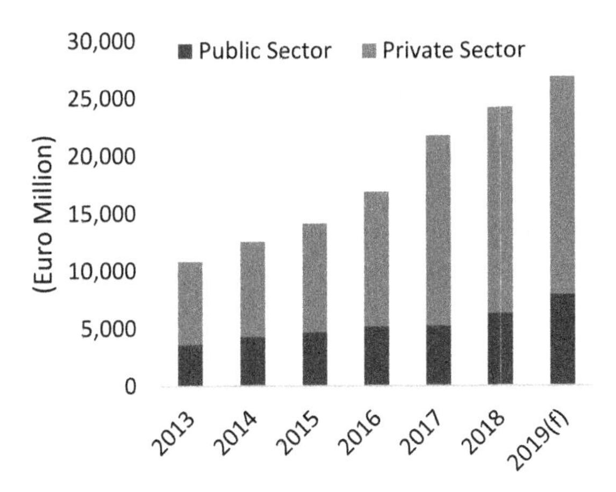

Figure 5.8 Investment in building and construction, Ireland
Source: Department of Public Expenditure and Reform, Government of Ireland (2020, June).

strategy to make Ireland a better country for all and to build a more resilient and sustainable future. The report states that 'Under this €116 billion plan, public capital investment will reach 4% of national income in 2020, placing Ireland well above the recent EU average of 2.9%'.

Prospects provides further visibility to the domestic and international construction industry on the sequencing of Ireland's priority infrastructure projects over the coming years. The report will also facilitate construction firms investing in their capacity and planning commercial bids for these major infrastructure priorities which are coming down track. One such project is the Limerick Regeneration Programme (see Figure 5.9).

Case study – Limerick Regeneration Programme, Ireland

The Limerick regeneration project, valued at €253 million, aimed at regenerating Limerick city through direct Government investment in physical, social and economic projects throughout Limerick. Fundamentally the regeneration package was designed to tackle crime, unemployment, and poor-quality housing.

The project at a glance:

Cost:

- €253 million euro on physical projects

plus

- €30 million euro on social regeneration projects
- €10 million euro on economic regeneration projects

Dates: Commenced in 2013, ongoing

Size:

- €10millioninvestment inroad projects
- €102 million investment in housing
- €12 million investment in community upgrading projects

Project funding:

The Irish government, European funding, the voluntary housing sector and the private sector.

Project benefits:

According to Limerick Council, the following benefits have been delivered:

- More than 300 jobs have been created.
- 16,871 participants in support services across 47 projects.
- Increased numbers using services at local community hubs.
- Two thirds of 1,500 homes earmarked for thermal upgrading either completed or plans in place.
- 30 projects increasing connectivity to the wider community.
- €12.2 million total employment income impact in 2015 up over a quarter in 2014.
- Increased attendance at school and other services and higher numbers from DEIS schools progressing to a further level.
- Almost 50 per cent of newly built homes are either completed or at a design stage.
- Safer communities working in partnership with communities, policing and CCTV; Youth crime has decreased significantly.

Project update:

> Limerick Regeneration, launched to fanfare 13 years ago, has 'failed' the communities it was tasked to help, Limerick's Mayor Daniel Butler has said. The projected €3bn project, drastically scaled down to €337m following the economic crash in 2008, and then in 2013 anticipating an average of €28m per annum, aimed to transform severely impoverished local authority housing estates.
>
> (Rayleigh, 2021)

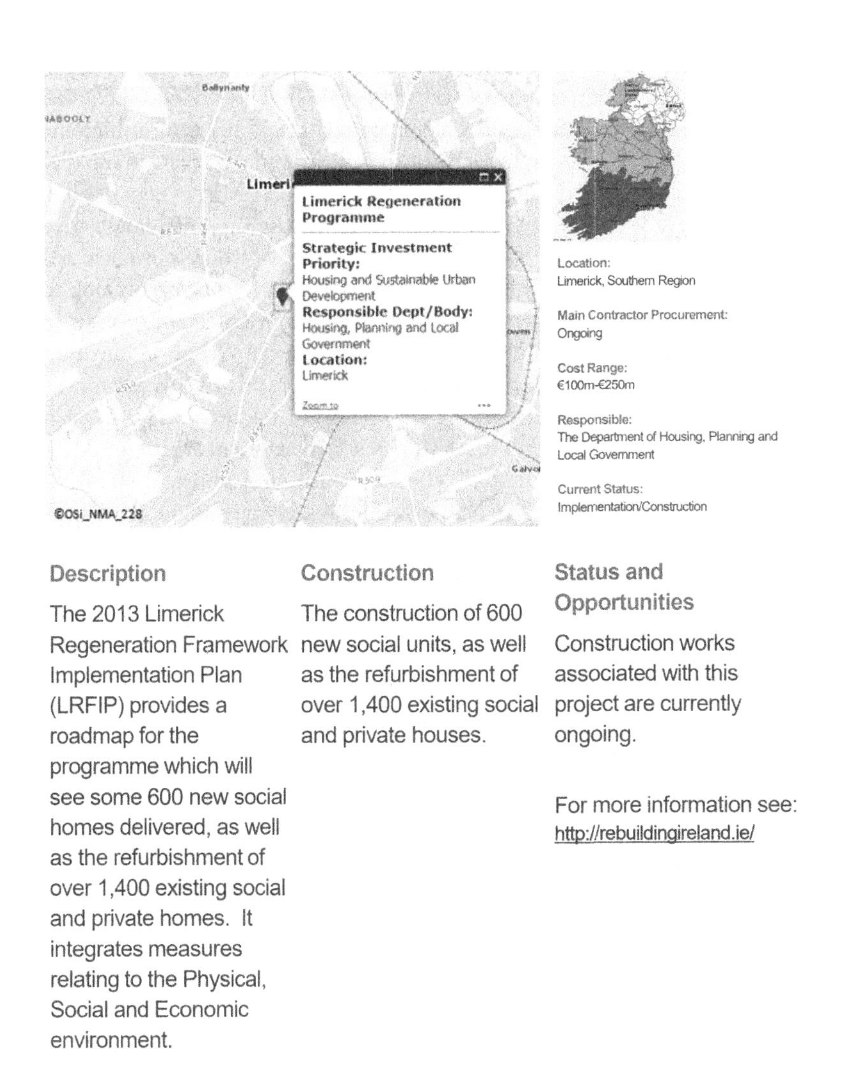

Figure 5.9 Limerick Regeneration Programme
Source: Investment Projects and Programmes Office, December (2019).

Developing economies

Developing nations require a sustained and significant investment in their infrastructure assets, which has been further exacerbated by the Covid-19 pandemic and the climate emergency. For example, Bhattacharya et al. (2015) found that investments of circa to $250 billion over a period of 15 years (2015–2030) are required for renewable energy (i.e., water, electricity, and transport).

As discussed above, the Global Infrastructure Hub (2021b) found that middle- and low-income countries attract only a quarter of the global private investment in infrastructure projects and saw a 28 per cent decline in private investment in 2020. Most of this investment occurs in the non-renewable energy and transport sectors. In 2010, private investment in infrastructure projects in middle- and low-income countries were around 0.4 per cent of GDP compared with 0.25 per cent in high-income countries. But, by 2020 private investment in infrastructure projects in middle- and low-income countries had declined significantly to around 0.11 per cent of GDP

From Figure 5.10 it can be seen that Asia was the most privately invested in region within the middle to low-income countries, whereas Africa was the least. However, Africa was the only region not to suffer negatively due to the pandemic and in fact was the only region apart from Eastern Europe to have increased private investment in infrastructure projects. Other regions have, in the past decade, witnessed a substantial decrease in private investment; Oceania, for example, has experienced a ~56 per cent decrease. There are many reasons for the substantial decreases in private investment in the middle- and low-income regions. These include the pandemic and political stability. Ultimately there exist significant barriers to private investment in infrastructure in these regions, which create greater disparity between developed and developing nations and necessitating greater public sector investment going forward.

Asia

A recent study of 25 of the developing member countries (DMCs) of the Asian Development Bank (ADB, 2017) identified that $204 billion of the $457 billion annual infrastructure gap would need private sector investment. This private

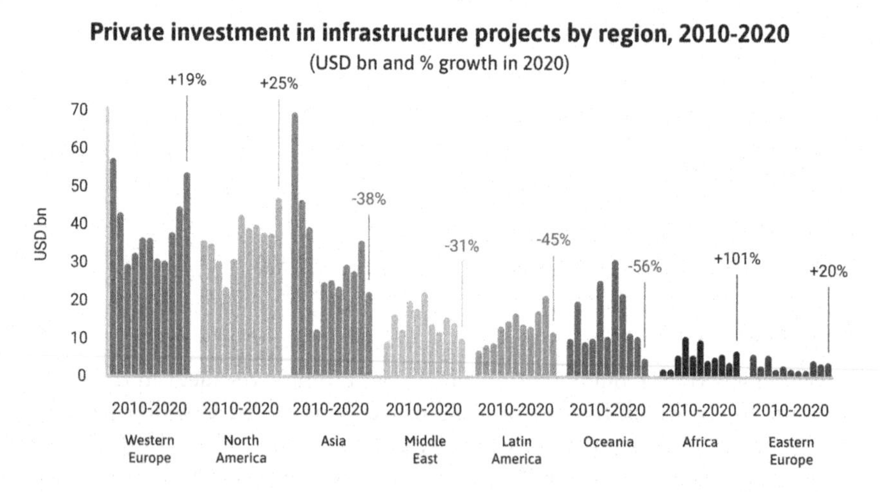

Figure 5.10 Private investment in infrastructure projects by region, 2010–2020 (US$ billion and percentage growth in 2020)

Source: Global Infrastructure Hub (2021b).

sector investment requirement is in addition to the 257 ADB approved loans over the period 2013–2016.

India

India has embraced PPP for the delivery of both economic and social infrastructure assets and is considered one the leading adopters of PPP in south Asia. McKinsey (2010) noted that the economy of India is projected to grow by 40 per cent by the end of 2030, which will also lead to an increase in Asia's infrastructure needs. In Figure 5.11 it can be seen that from 1990–2016, India had committed over US$314 billion to infrastructure investment across the transport, energy, water and social infrastructure sectors. Of these PPP projects, 18–19 per cent were funded by foreign private investors. Most of these investment transactions were related to BOT and DBFOT concession contracts, either for new projects (i.e., greenfield stage) or the development of existing assets (i.e., brownfield stage). The driver behind this strategy was to involve a maximum amount of local and global foreign investors to accelerate the infrastructure investment rate (ADB, 2017).

And, however, although there was a substantial increase in PPP investments of US$48.9 billion in 2010, India has witnessed a significant decline in PPP investment thereafter with investments of US$4.12 billion in 2015 and US$2.01 billion in 2016 (ADB, 2017). The biggest challenge to securing private investment is the Indian regulatory system, particular land issues, that create delays and increased risk for potential investors. As a result of these barriers, infrastructure investments (between fiscals 2013 and 2019) were predominantly made by the public sector (70 per cent). Furthermore 95 per cent of these infrastructure investments were in economic sectors including power, roads and bridges,

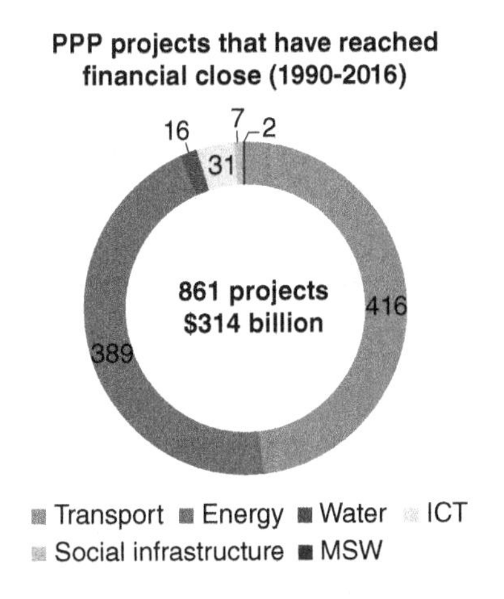

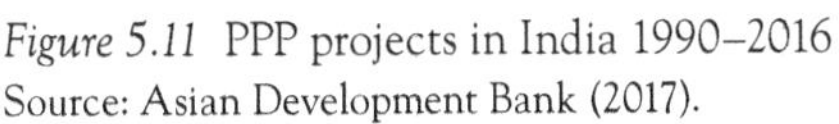

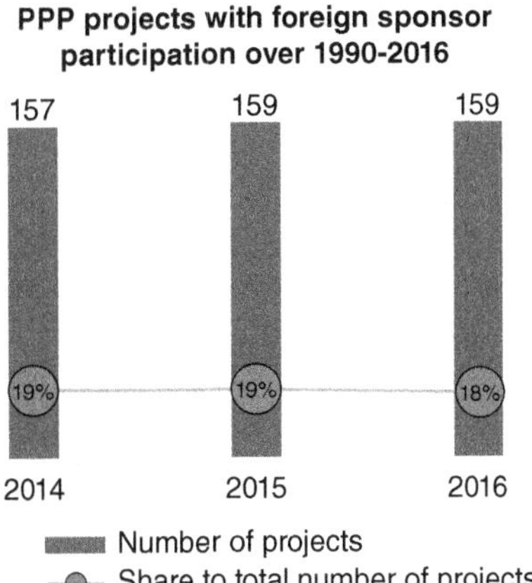

Figure 5.11 PPP projects in India 1990–2016
Source: Asian Development Bank (2017).

telecommunications, railways, irrigation and urban infrastructure (Government of India, 2020a).

Central government is responsible for producing PPP guidelines which include formulation, appraisal and approval of PPP projects. Furthermore, this guidance identifies the government's favoured PPP approaches which include: Management contracts and BOT (Build–Operate–Transfer) and its variants– Build–Lease–Transfer (BLT), Design–Build–Finance–Operate–Transfer (DBFOT) and Operate–Maintain–Transfer (OMT). Under these PPP models, funding is through a project finance arrangement whereby lenders rely exclusively on the cash flow expected to be generated by the project to recover loans and earn a return on their investments.

	2014	2015	2016
Project funding structure	Applicants should have minimum net worth equivalent to 25% of estimated capital cost of projects for which bids shall be invited.	Applicants should have minimum net worth equivalent to 25% of estimated capital cost of projects for which bids shall be invited.	Applicants should have minimum net worth equivalent to 25% of estimated capital cost of projects for which bids shall be invited.
Project capital investment size	None	None	None

Figure 5.12 PPP project funding restrictions
Source: ADB (2017).

In 2016, the Indian Department for Economic Affairs produced a *PPP Guide for Practitioners* aimed at assisting project authorities in developing their capacities for PPPs, including procurement, contract management, etc. The guidance also included details on funding restrictions and advises that 25 per cent of the capital cost of the project is required by applicants if PPP is to be considered (see Figure 5.12). The leading deals within the transport sector in 2015 were the US$26,211 million Fagne–Maharashtra/Gujarat Border Road Upgrade Package III Project.

Redevelopment of the Anand Vihar Railway Station by Indian Railway Station Development Corporation Limited is part of the continuous process of augmenting and improving amenities keeping in view the Indian Government's goal of 'World Class' planning by 2038. For this, Indian Railways have identified five major railway stations to be developed to World Class standards, including the Anand Vihar Railway Station in East Delhi. Anand Vihar Railway Station is going to be a big multi modal transit hub with five modes of transport culminating at a single location: 1. Indian Railways Station 2. Existing Delhi Metro Phase I station 3. Anand Vihar ISBT 4. Proposed Delhi Metro Phase III station 5. Proposed Delhi Meerut high speed Regional Rail Transit Corridor station. The station has been planned for development in two phases: Phase 1 to be completed by 2025 and Phase 2 by 2038 (Ministry of Environment, Forest and Climate Change, 2017). Further details of the project are given in the case study below.

Case Study – Redevelopment of Anand Vihar Railway Station, Delhi

Indian Railway Stations Development Corporation Limited's (IRSDC) ambitious station redevelopment plan for the Anand Vihar railway station will offer travellers a seamless travel experience and is a big step taken by IRSDC. Anand Vihar serves as a key junction joining the whole of the eastern region of Delhi, whilst Delhi serves as a focal point and key junction of connectivity to the rest of the country.

The project at a glance:

Cost: US$28.1 million

Dates: Pre bid meeting, 2020, ongoing.

Project goals:

- The total area available for site development is 5,68,145.00 m^2 (56.8 ha).
- Redevelopment of the main station building with construction of facilities such asa hotel, station retail and a commercial office.
- The parking shall be provided in the basement and 34,000 m2 will be maintained as garden and lawns.
- The railway stations will have multi-modal integration and will be connected to bus terminals and city metro stations.

Project funding:

RSDC is a special purpose vehicle (SPV) under the Ministry of Railways and has been entrusted with the task of redeveloping railway stations, for providing state-of-the-art passenger facilities. Anand Vihar and Bijwasan railway stations will be awarded on the EPC or Engineering, Procurement and Construction mode. The Ministry called for interested developers and funders to privately finance the scheme. Interested bidders included Dani Realty, Bharti Realty, Larsen, and Toubro, GMR, Eldeco, Godrej, R Cube, Kalpataru Power Transmission, and R Retail Ventures.

Project benefits:

Reduce city decongestion by distributing traffic, whilst ensuring that the land of the railway station is used for multiple purposes and that there is a round-the-clock city centre at the railway station complex.

Delivering an environment friendly station through the movement of pedestrian and non-motorised transport (NMT).

To summarise, as mentioned above, the Asian Development Bank reported back in 2017 that India had witnessed a significant decline in PPP investment since 2014, citing, among other reasons, unfavourable investor market conditions, delays in obtaining environmental and forest clearances for the private sector and regulatory challenges. Yet, more recently, the Indian Government (2020, Volume I) has stated that 'We expect India's GDP to recover in the five years beginning fiscal 2021 (2020–21 to 2024–25)' but it warns that it would need to 'spend $4.51 trillion on infrastructure by 2030 to realise the vision of a $5 trillion economy by 2025, and to continue on an escalated trajectory until 2030'.

People's Republic of China (PRC)

The centre of global infrastructure projection has now shifted to China, with McKinsey forecasting the infrastructure investment needed in China to be approximately US$16 trillion through to 2030. In 2019, China invested over US$120 billion in its ten largest infrastructure projects by value, despite its economy showing signs of slowing growth (Haran, 2021). In terms of source of capital, public funding continued to provide the majority of that investment. During 2017, there were 14 PPP projects in various stages of preparation and 13 under procurement. Most of these projects are in the renewable energy sector. At 1.4 billion people, China's population is the largest in the world, China could nonetheless potentially do more to innovate and export solutions to the world, for instance helping to define global digital governance, and to fill the world's estimated $350 billion annual infrastructure investment gap.

PPP has existed in China for almost 40 years, but only more recently has the Government sought to increase its use. In 2015, the HM Treasury developed national guidance on the implementation of PPP in China and the use of foreign investment to deliver national infrastructure programmes. Whilst there is no legislation on the formal adoption of PPP and the use of foreign investment, it is highly anticipated that legislation will be introduced to facilitate greater use of PPP. In April 2022, the Ministry of Finance issued statement concerning improving the quality of new projects and ensuring that the PPP model is more standardised and sustainable. Furthermore, the performance management module of Ministry of Finances PPP information platform has been officially launched. It has been put into trial operation in eight provinces and cities including Hebei, Shandong and Guizhou province, so as 'to further improve the intelligent and professional level of performance management by using information-based means' (China State Finance, 2022).

As can be seen from Figure 5.13, most of the PPP projects in China from 1990–2016 were in the water, energy, and transportation sectors. The projects could avail themselves of several different PPP models, including:

- operation and maintenance (O&M),
- management contract,
- BOT, build–operate–own, transfer–operate–transfer, and
- rehabilitate–operate–transfer (ROT).

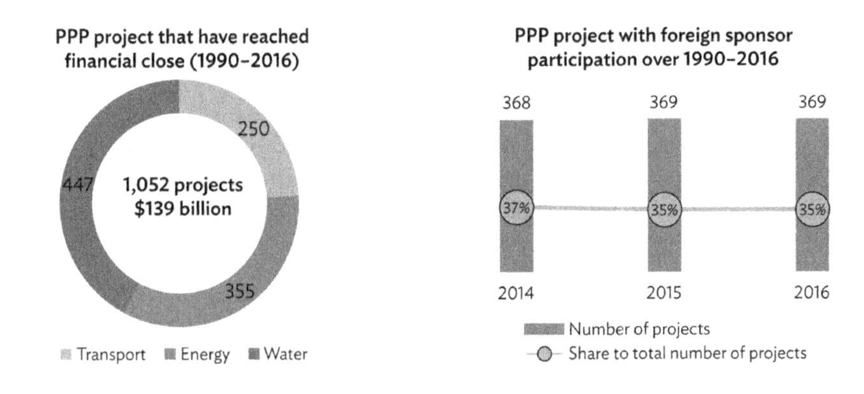

Figure 5.13 PPP projects in the People's Republic of China
Source: Asian Development Bank (2017).

	2014	2015	2016		2014	2015	2016
Transport				**Energy**			
Roads	✓	✓	✓	Power generation	✓	✓	✓
Railways	✓	✓	✓	Power transmission	✓	✓	✓
Ports	✓	✓	✓	Power distribution	✓	✓	✓
Airports	✓	✓	✓	Oil and gas	✓	✓	✓
Water and wastewater	✓	✓	✓	**ICT**	✓	✓	✓
Municipal solid waste	✓	✓	✓	**Social Infrastructure**	✓	✓	✓

Figure 5.14 Key infrastructure eligible sectors
Source: Asian Development Bank (2017).

The PRC had identified all economic and social infrastructure sectors eligible for PPP and issued guidance on how PPP within these sectors could deliver value for money. As can been seen in Figure 5.14, the key infrastructure sectors identified by the PRC that were eligible for PPP are across the spectrum for both economic and social infrastructure.

In recent years, the number of PPPs in China has soared and by October 2017 the government had initiated 14,220 projects with an expected total capital value of 17.8 trillion renminbi (RMB) (~US$2.6 trillion) (Tan and Zhao, 2019). Even more recently, in the first 5 months of 2022, China saw 165 new PPP projects registered. Of these, 51 were related to urban infrastructure (The State Council, PRC, 2021).

The Ministry of Finance (MOF) in China has assumed responsibility for producing guidance on the use of PPP, but it is facing significant challenges in procuring greater infrastructure investment, most notably due to the lack of a legislative framework for attracting private investors. However, with investment plans worth more than US$1.6 trillion, the PRC's expanding PPP programme has emerged as an important investment opportunity for pension funds. Hence 'Pension funds could become an important source of capital for PPPs alongside other institutional investors', which in turn will help to sustain their financial viability for an aging population (Asian Development Bank, 2021).

The various types of projects utilising PPP include energy, transportation, water conservancy, environmental protection, agriculture, forestry. One such project is the Jichang river enhancement project.

Case Study – Pollution Elimination and Environment Enhancement of Huangxiao River and Jichang River (Phase II), Wuhan, China

The Jichang River enhancement project is designed to eliminate the pollution and enhance the environment for two rivers in Wuhan Municipality–the Huangxiao River and Jichang River.

Project at a glance:

Cost: NY 4.775 billion Yuan, i.e. about US$734.6 million.

Dates: Procurement stage 2018, with an anticipated construction phase of 2.5 years and 15-year operation and maintenance.

Project goals: To eliminate the pollution and enhance the environment for the two rivers, Huangxiao and Jichang.

Project funding:

To be developed with PPP model, to include, Equity Debt, Private Sector Construction Operations Debt, Concession Debt, and Green Finance Initiative for a contract period of 16–20 years.

Project benefits:

To prioritise the protection and restoration of original ecosystems in the urban area, such as rivers, lakes, and wetlands, and to rebuild a harmonious relationship among people, water, and city.

Africa

Nigeria

There is a huge variance in terms of investment rates (deal values and numbers) between the African regions. Nigeria leads Africa in relation to a viable PPP project pipeline. This stems back to 2005 when the national government endorsed and pursued PPP investment through the approval of the Infrastructure Concession Regulatory Commission Act in 2005.

The Infrastructure Concession Regulatory Commission (ICRC) is still Nigeria's main PPP unit, with the key objective to foster private investment in the country's national infrastructure. The ICRC assists the federal government and its ministries and development agencies in implementing and establishing effective PPP processes. In 2019, investment in PPP projects reached an amount of US$1.6 billion, after US$29.6 billion in 2018 and US$3 billion in 2017. Collectively this represents a total of 50 active PPP projects, 19 of which are in construction and 31 projects in operation. As indicated in Figure 5.15, the transport sector leads the PPP market in the country in both the number of projects and investment volumes (65 PPP projects and US$31.2 billion). Energy and Social & Health share the remaining part of the market.

The closest competitor in terms of PPP to Nigeria is Ghana with 44 projects in the pipeline (US$3.9 billion) and 11 active projects (US$8.1 billion). Kenya, the Democratic Republic of the Congo, and Mozambique are also active in the PPP market. These countries have a total of 15 active projects and 17 PPP projects in the pipeline. However, there is huge variance in terms of investment rates (deal values and numbers) between the African regions (see Figure 5.16).

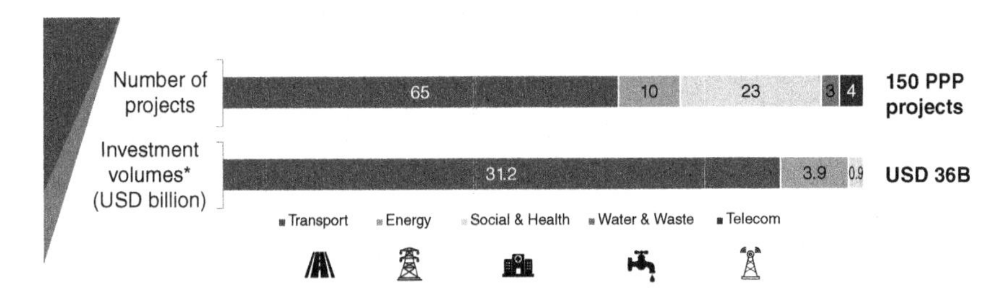

Figure 5.15 PPP project portfolio by sector, Nigeria
Source: InfraPPP World Project Database (2019).

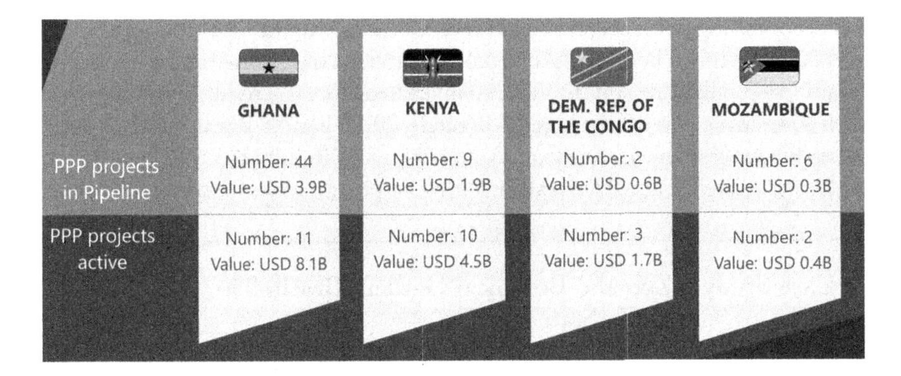

Figure 5.16 Top regional PPP competitors, Africa
Source: InfraPPP World Project Database (2019).

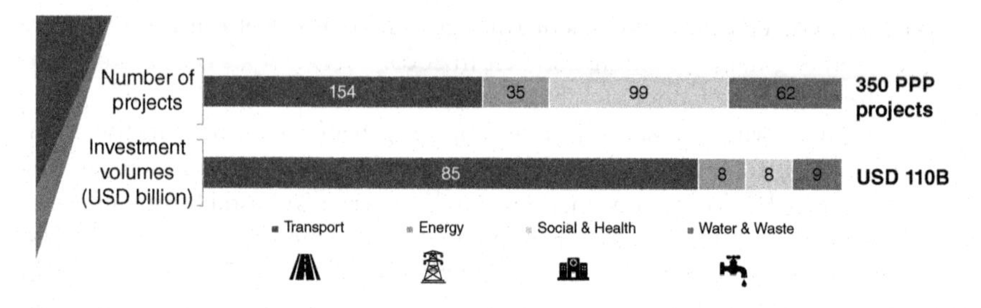

Figure 5.17 PPP project portfolio by sector, Brazil
Source: InfraPPP (2019).

South America

Brazil

The environment for PPP projects in Latin America and the Caribbean has evolved positively in the last few years, with 16 countries in the region creating PPP-dedicated agencies which provide technical support and oversee private participation in infrastructure, demonstrating political support for private investment.

Brazil has the largest PPP market by far in Latin America, representing 50 per cent of the total number of projects and 45 per cent of the total investment volume in the continent. Government investment in PPP is facilitated through the Investment Partnerships Programme in 2016 and the Project Support Fund in 2017.

As can be seen in Figure 5.17, the transport sector dominates the PPP market in Brazil in both number of projects and investment volumes (154 PPP projects and US$85 billion). Social & Health, Water, and Energy share the remaining investment on PPP projects (196 PPP projects and US$25 billion). There are also an additional 195 PPP projects in the pipeline representing a further investment of US$42 billion.

Brazil has attracted global private investors and has acquired a strong position at the region. Foreign direct investment into Brazil boomed between 2009–2011, but has been slowing down ever since. However, according to the World Investment Report 2021, published by UNCTAD (2021), FDI inflows have decreased by 62 per cent, from US$65 billion in 2019 to US$25 billion in 2020. As a result, Brazil have seen their sovereign credit ratings lowered (Reuters, 2018). In 2018 Brazil launched its privatisation programme at securing private investment in infrastructure, however the pandemic created a significant stall and only two privatisations have been completed.

Case Study – Zoo and Botanical Garden, Brazil, São Paulo State

Investment in the Zoo and Botanical Gardens, São Paulo, will comprise of investments in the visiting areas, the green areas, enhancing conservation, environmental education, and leisure. This investment will build on the success of the Zoo, which has annual visitation of 1 million people, and the Botanical Garden, which has annual visitation of 270,000 people.

Project at a glance:

Cost: BRL $359 million

Dates: Project bid 2021, term of 30 years

Project size:

- The São Paulo Zoo is the largest zoo in Brazil. With 824,529 m2 (82.45 hectares [203.7 acres]) of space in what was originally the Atlantic Forest, the Zoo is south of the city of São Paulo.
- The Botanical Garden spans a 164.45-hectare area.

Project funding:

Reserva Paulista Consortium proposal registered the highest grant value (US$19.5 million) to manage public spaces for 30 years.

Project benefits:

At the Zoo, the concessionaire will have to promote more immersion in nature and in the enclosures for animal welfare. There are minimum investments planned to build more modern, broad, and integrated habitats, in addition to monitoring animal health indicators.

Research and conservation of species threatened with extinction will continue under the responsibility of the State Government for the duration of the concession.

The project also foresees that the winning company should grant free tickets for children up to 4 years old and for students and teachers of early childhood education, elementary and high school in the public school system, on specific days. The right to pay half-price is also guaranteed.

At the Botanical Garden, the project's idea is to increase public use with the implementation of environmental education programs, new spaces for leisure and culture, food, and more accessibility, integrating with research activities.

Summary

There is a significant demand for increased infrastructure investment across the globe. It is now accepted that infrastructure investment is required to support sustainable economic and social development. In the last decade, wholesale inadequacies in national government budgets have promoted the increase in private capital to support essential infrastructure.

However, there are significant international challenges to the investment of infrastructure. In the context of developing countries, there are a considerable number of challenges that inhibit private investor opportunities. These include social unrest, economic instability, political unrest, lack of knowledge and experience (Inderst and Stewart, 2014). Yet these developing countries present an opportunity to investors of infrastructure as the demand in these markets is immense, particularly in the energy sector. Developed countries also have inherent challenges in providing sustainable infrastructure that commits to delivering on reducing carbon emissions. This will require all nations to collaborate and innovate.

References

Aschauer, A. (1989). "Is public expenditure productive?", *Journal of Monetary Economics*, vol. 23, pp. 177–200.

Aschauer, D. (1990a). "Why is infrastructure important?" Federal Reserve Bank of Boston, 21–68.

Aschauer, A. (1990b). "Is government spending stimulative?", *Contemporary Economic Policy*, vol. 8, no. 4, pp. 30–46.

Asian Development Bank (2017). *Public–Private Partnership Monitor*. Asian Development Bank. Available at: https://www.infrapppworld.com/report/public-private-partnership-monitor

Asian Development Bank (2021, December). *People's Republic of China: Mobilizing Pension Fund Financing for Public–Private Partnerships*. TCR Evaluation Report. Asian Development Bank. Available at: https://www.adb.org/sites/default/files/evaluation-document/762341/files/tcrv-9163.pdf

Bhattacharya, A., Oppenheim, J. and Stern, N. (2015). "Driving sustainable development through better infrastructure: Key elements of a transformation program". Brookings Global Working Paper Series. Available at: https://www.brookings.edu/research/driving-sustainable-development-through-better-infrastructure-key-elements-of-a-transformation-program/

Bhattacharya A., et al. (2016). "Framework for assessing the role of sustainable infrastructure", Brookings Institution, Washington DC.

Catalano, G. and Sartori, D. (2013). "Infrastructure investment long term contribution: Economic development and wellbeing", Working Papers 201301, CSIL Centre for Industrial Studies, Milan.

China State Finance (2022, February). "Performance management: Strengthen whole-process performance management and promote high-quality development of PPP". Press release. Available at: http://www.cpppc.org/en/szyw/999349.jhtml

Construction Sector Group (2020). *Project Ireland 2040*. Available at:https://www.gov.ie/en/campaigns/09022006-project-ireland-2040/

De la Fuente, A. and Estache, A. (2004). "Infrastructure productivity and growth: A quick survey", WBIGF, Mimeo, Washington DC.

Department of Public Expenditure and Reform (2020, July). *Build 2020: Construction Sector Performance and Capacity*. Government of Ireland. Available at: https://www.gov.ie/en/publication/c19a5-build-2020-construction-sector-performance-and-capacity/

Department of Public Expenditure and Reform (2020, June). *Construction Sector Group – Building Innovation*. Government of Ireland. Available at: https://www.gov.ie/en/publication/827c7-construction-sector-group-building-innovation/

Department of Public Expenditure and Reform, Investment Projects and Programmes Office (2019). *Project Ireland 2040. Prospects – Ireland's Pipeline of Major Infrastructure Projects*. Department of Public Expenditure and Reform, Ireland. Available at: https://www.gov.ie/en/collection/580a9d-project-2040-documents/

Fakhoury, I. N. (2022, May). "New data shows private investment lends a hand as public debt looms large". World Bank Blogs, 3 May. Available at: https://www.gov.ie/en/collection/580a9d-project-2040-documents/

Global Infrastructure Hub (2018). *PPP Project – Pollution Elimination and Environment Enhancement of Huangxiao River and Jichang River,(Phase II)*. Project Pipeline. Available at: https://pipeline.gihub.org/Project/ProjectDetails/521

Global Infrastructure Hub (2021a). "Forecasting infrastructure investment needs and gaps". Available at: https://outlook.gihub.org/

Global Infrastructure Hub (2021b). "Private investment in infrastructure". Available at: https://cdn.gihub.org/umbraco/media/4302/gihub_infrastructuremonitor2021_private_investment_in_infrastructure.pdf

Global Infrastructure Hub (2021c). "Zoo and Botanical Garden (Sao Paulo State)". Available at: https://pipeline.gihub.org/Project/ProjectDetails/521

Government of India, Department of Economic Affairs (2016, April). *PPP Guide for Practitioners*. Available at: https://www.pppinindia.gov.in/documents/20181/33749/PPP+Guide+for+Practitioners/

Government of India (2020, April). *National Infrastructure Pipeline Volume I*. Report of the Task Force Department of Economic Affairs, Ministry of Finance, Government of India. Available at: https://www.pppinindia.gov.in/documents/20181/55954/Report+of+the+Task+Force+National+Infrastructure+Pipeline+%28NIP%29+-+volume-i_0.pdf/5f0c7cd5-3409-4267-8010-01ac5c2ab98e

Government of India (2020, April). *National Infrastructure Pipeline Volume II*. Report of the Task Force Department of Economic Affairs, Ministry of Finance, Government of India. Available at: https://www.pppinindia.gov.in/documents/20181/55954/Report+of+the+Task+Force+National+Infrastructure+Pipeline+%28NIP%29+-+volume-ii_0.pdf/2a4cb70c-2dc8-467a-a0ac-a7111537d633

Grimsey, D. and Lewis, M.K. (2002). "Evaluating the risks of public private partnerships for infrastructure projects", *International Journal of Project Management*, vol. 20, no. 2, pp. 107–118.

Haran, M. (2021, July). "How does China pay for her infrastructure? – An introduction". RICS World Environmental Forum, 8 July. Available at: https://www.rics.org/uk/wbef/megatrends/markets-geopolitics/case-study-how-does-china-pay-for-her-infrastructure--an-introduction/

HM Treasury (2015, March). *Supporting the Development of PPPs in the People's Republic of China: An International Perspective*. Available at: https://assets.publishing.service.gov.uk/government/uploads/system/uploads/attachment_data/file/438892/IUK_Report_1_Chinese_PPP_Enabling_Environment__executive__FINAL.pdf

Inderst, G. and Stewart, F. (2014). *Institutional Investment in Infrastructure in Emerging Markets and Developing Economies*. World Bank Group, Washington, Public–Private Infrastructure Advisory Facility (PPIAF), Series 91307. May 2014.

InfraPPP (2020). *Brazil PPP Market 2020*. InfraPPP Reports. Available at: https://www.infrapppworld.com/report/brazil-ppp-market-2019

InfraPPP (2022, May). "Real Time Tools". Available at: https://www.infrapppworld.com/real-time (accessed 26 May).

InfraPPP World Project Database (2019). "Top PPP players in USA 2018/19". Available at: https://www.infrapppworld.com/report/usa-ppp-market-2019

Limerick City & County Council (2016, November). *Limerick Regeneration Framework Implementation Plan Review*. Available at: https://www.limerick.ie/council/services/housing/regeneration/limerick-regeneration-framework-implementation-plan-review

McKinsey (2021a). "A sustainable, inclusive and growing future for the United States". McKinsey. Availableat: https://www.mckinsey.com/featured-insights/sustainable-inclusive-growth/a-sustainable-inclusive-and-growing-future-for-the-united-states

McKinsey (2021b). "Infrastructure and Investment Jobs Act". McKinsey. Available: https://www.mckinsey.com/industries/public-and-social-sector/our-insights/the-us-infrastructure-investment-and-jobs-act-breaking-it-down

McKinsey Global Institute (2010). "India's urban awakening: Building inclusive cities, sustaining economic growth". McKinsey & Company. Available at: https://www.mckinsey.com/featured-insights/urbanization/urban-awakening-in-india

McKinsey Global Institute (2013). "Infrastructure productivity: How to save $1 trillion a year". McKinsey Global Institute. Available at: https://www.mckinsey.com/business-functions/operations/our-insights/infrastructure-productivity

Ministry of Environment, Forest and Climate Change (2017). *Proposed Redevelopment of Anand Vihar Railway Station, Anand Vihar, East – Conceptual Plan*. Online Submission & Monitoring of Environmental & CRZ Clearances, Government of India. Available at: http://environmentclearance.nic.in/writereaddata/Online/TOR/23_Feb_2017_1455434331XJT9HJKAnnexure-PFR.pdf

New York State (2021, December). "Govenor Horchul announces major ground breaking in the Port Authority's JFK transformation project". 15 December. Available at: https://www.airport-technology.com/analysis/jfk-expansion-project/

OECD (2019). "Business insights on emerging markets 2019". OECD Emerging Markets Network, OECD Development Centre, Paris. Available at: http://www.oecd.org/dev/oecdemnet.htm

PwC (2014). "Capital projects and infrastructure spending, outlook to 2025". PwC. Available at: https://www.pwc.co.za/en/assets/pdf/capital-projects-and-infrastructure.pdf

Rayleigh, D. (2021). "Mayor of Limerick: City's €331m regeneration has 'failed'". the journal. ie, 18 September. Available at: https://www.thejournal.ie/limerick-regeneration-has-failed-mayor-says-5552586-Sep2021/

Reuters (2018). "S&P cuts Brazil credit rating as pension reform doubts grow". Reuters website, 12 January. Available at: https://www.reuters.com/article/brazil-sovereign-downgrade/sp-cuts-brazil-credit-rating-as-pension-reform-doubtsgrow-idUSL1N1P628Y

Rutherford, A. (ed.). (2002). *Routledge Dictionary of Economics*, 2nd edition(London and New York: Routledge).

Tan, J. and Zhao, J. Z. (2019). "The rise of public–private partnerships in China: An effective financing approach for infrastructure investment?"*Public Administration Review*, vol. 79, no. 4, pp. 514–518. Available at: https://www.webssa.net/files/tan2019.pdf

The Department of Public and Review (2020). *Construction Sector Group. Project Ireland 2040*. Government of Ireland. Available at: https://www.gov.ie/en/campaigns/09022006-project-ireland-2040/

The State Council, The People's Republic of China (2021). "China adds 165 new PPP projects in January–May", 28 June 2021. Available at: http://english.www.gov.cn/statecouncil/ministries/202106/28/content_WS60d9022fc6d0df57f98dbfb1.html#:~:text=BEIJING%20%E2%80%94%20China%20saw%20165%20new, Development%20and%0Reform%20Commission%20shows

The Whitehouse Press Office (2021). "Fact sheet: The Bipartisan Infrastructure Deal". The Whitehouse Press Office. Available at: https://www.whitehouse.gov/briefing-room/statements-releases/2021/11/06/fact-sheet-the-bipartisan-infrastructure-deal/

Tunnicliffe, A. (2020, May). "Departure: Inside the ambitious JFK Terminal 4 redevelopment project". Airport Technology, 28 May. Available at: https://www.airport-technology.com/analysis/jfk-expansion-project/

UNCTAD (2021). *World Investment Report 2021 – Investing in Sustainable Recovery*. Available at: https://unctad.org/webflyer/world-investment-report-2021

World Bank (2014). "Prioritizing projects to enhance development impact". World Bank Group. Available at: http://www.g20.utoronto.ca/2014/6%20Prioritizing%20Projects%20to%20Enhance%20Development%20Impact.pd

6 Challenges to infrastructure investment

Introduction

This final chapter will explore and contextualise the key challenges to greater investment in infrastructure. The key barriers to investment are economic, regulatory and political. This chapter offers insights into how these investment challenges can be mitigated and the new measures that governments are actioning to secure infrastructure investment that delivers a sustainable, resilient community for all.

Learning outcomes

In this chapter you will be able to:

1 Appraise the key economic and political barriers to infrastructure investment.
2 Identify the key measures governments are taking to overcome these investment challenges.
3 Appreciate the evolving challenges in the delivery of infrastructure.

Key challenges to greater investment in infrastructure

Globally, governments continue to wrestle with the consequences of underinvestment in infrastructure. For any region to be successful, modern and world class infrastructure needs to be in place. According to Sherraden (2011), 'The infrastructure deficit is the result of a steady decline in government infrastructure spending, combined with a steady increase in the cost of building additional infrastructure'. As stated, bluntly, by Haider (2017), 'the world is facing a severe infrastructure deficit. The reasons for this severe infrastructure deficit are multifaceted, complex, and evolving'. Hence, this last chapter seeks to provide readers with a clear understanding of the salient factors that hinder greater infrastructure investment. According to HM Treasury (November 2020), in the UK there are varying reasons for the underinvestment in infrastructure, identified as:

- Insufficient long-term planning by successive Governments;
- Inefficient stop–start public sector investment;

DOI: 10.1201/9781003171805-6

- Inefficient funding for regions outside of London;
- Slow adoption of new technology;
- Policy uncertainty that undermines private investment;
- Project delivery plagued by delays and cost overruns.

Fundamentally the salient factors impacting on governments' ability to deliver greater investment in infrastructure are economic and political. A report from McKinsey and Company in 2017 (Woetzel et al., 2017) found that up to 38 per cent of global infrastructure investment is not spent effectively because of bottlenecks, lack of innovation and market failures. Governments could reduce spending by employing fact-based project selection, streamlining the delivery and the optimisation of operations and via maintenance of existing infrastructure (McKinsey, 2020).

Understanding the economic barriers to infrastructure investment

Infrastructure investment is of vital importance to both developed and developing economies. There is a vast challenge to finance infrastructure to counter the global deficit. Current trends have predicted that there will be an investment need to spend US$4.6 trillion per year in the years to 2040 (Carroll, 2022). This would also have to increase even further to meet the United Nations' sustainable development goals. Furthermore, the cost to repair and maintain existing infrastructure has also contributed to the fast-growing deficit. In the USA, for example, of its 617, 000 bridges, almost 42 per cent are more than 50 years old, causing real concern about public safety and the economic effects of repair and reduced mobility. This has prompted a policy change to overcome years of underinvestment, with the current focus now on retrofitting existing infrastructure rather than building (ASCE, 2021).

Economic barriers to infrastructure investment centre around limitations in the public finance available and therefore alternative sources must be sought. Market fragmentation can also hinder infrastructure investment. The European Investment Bank (2021) highlights that this can impact investment in two ways: By reducing the size of the potentially accessible market and by obstructing cross-border competition. A country's economic performance can also impact on infrastructure investment. If productivity levels are low, domestic and foreign capital is less likely to be attracted to the region, in turn stunting socioeconomic growth.

The sustained lack of public capital investment in infrastructure has now resulted in a widening infrastructure investment gap, which has created an even greater challenge for governments wishing to invest in essential infrastructure. The sustained underinvestment in infrastructure is due to many social, economic, and political factors including the economic shock of the 2020/21 pandemic and the long-term consequences of the global financial crisis of 2008; the demand for more and better essential infrastructure for increasing and ageing populations; and political instability of countries.

These factors have increased uncertainties both for governments and for investors–uncertainty that has stagnated infrastructure investment and decisions to invest. Government and municipalities currently seem to be unable to mitigate these uncertainties and to collaborate with the private sector to deliver an increased sustained infrastructure investment. Global research by the Coalition for Urban Transitions (2017) identified the following economic barriers to infrastructure investment:

1. Lack of upfront public capital

   Government lacks the upfront capital to fund its investment priorities. Since the financial crisis, there has been a slowdown in economic growth in the UK and other developed economies, resulting in affordability pressures on consumers and taxpayers. These risks can come in the form of unforeseen tariff changes, directives to increase competition, and a weakening of the independence and transparency of regulatory institutions.
2. Institutional inertia

   The difficulty of changing investment patterns due to institutional, governance, and contractual/financial features present in the market. Post financial crisis (2008), governments have increased regulation on all investments due to the housing market scandal. Regulations have been introduced to promote competitive markets and to stop monopoly firms controlling markets. However, these can also act as barriers to investment. Internationally, governments aim to attract inward investment through infrastructure development and yet due to this tightening of regulations some investors will not achieve a desired return on capital and therefore will not invest. Kemna (2015) further suggests that a failure to appropriately regulate could inhibit long-term investments that have the potential to be widely beneficial to economic growth. Kemna gives a pinpoint example of poor regulation where pension funds, that pose minimal systematic risk to financial markets, must set aside collateral similar to hedge funds, resulting in major losses in long-term investment opportunity.
3. Institutional capacity

   National, regional and municipal governments cannot initiate projects or act as bankable counter parties due to structural, technical, and skills limitations. In emerging countries, the economic situation tends to be volatile and unstable, increasing the level of investment risk due to immature institutional capacity.
4. Risk

   Infrastructure investors and operators can suffer monetary losses due to an unexpected decrease in capital returns or default from a project counter party to meet contracted obligations (OECD, 2015). Infrastructure investments can involve high levels of risk which can achieve greater reward; however, the probability of incurring a loss also increases. Bitsch et al. (2010) highlighted that in general infrastructure projects require a significantly larger amount of capital than non-infrastructure projects to fund and finance and therefore there is more risk to larger volumes of capital. Risk is further increased because of the broad timeframe of infrastructure investments.

5 Low returns

Investors forecast that an investment in infrastructure will generate insufficient returns, e.g. through debt repayments, asset appreciation or income streams, as a return on equity relative to other sectors and asset classes.

6 Imperfect information

Investors possess insufficient information on the opportunities that exist and how worthwhile an opportunity may be. Predicting how much investment is needed is complex. Several organisations have, however, made progress in this area, developing various analytical tools and commentary to compare spending requirement across nations and subsequently determine future investment need.

Collectively these barriers represent a public and private sector immaturity around the economic long-term investment decisions. According to the WBEF (2020) survey, governments that had developed a planned infrastructure pipeline prior to the pandemic have been better placed to respond to the crisis. Regrettably, across the G20, few countries have pipelines that link to a longer-term vision. This short-sightedness could become more damaging still if, in the rush to kick-start the global economy, shovel-readiness becomes the key criterion in funding decisions.

Another key economic reason for the global infrastructure deficit has been the slow and long international recovery which followed the global financial crash as nations significantly reduced investment during this recovery phase;an approach that some experts would argue has only perpetuated the pain. Since the financial crisis, the financial resources available to governments have become tighter. Despite the failings of PFI, including contractual non-compliance, infrastructure investment requires long-term arrangements with investors. PPP arrangements provide for these long-term investments. However, investors, both debt and equity, should recognise that over time there can be shifts in Government policy that may have a negative or positive impact on a PPP contract.

Government seeks the most cost-effective way of providing infrastructure within the prevailing political environment. Over time this can involve working with the private sector and providing creditor-friendly systems to a greater or lesser extent. To attract private investors into the infrastructure financial market it is crucial that investors are attracted to the investment opportunity through managed risks and are rewarded accordingly with sufficient return on investment. These investment opportunities must be bolstered by governments who are prepared to provide low-risk debt repayments.

Commercial banks are the largest potential source of private finance for infrastructure projects and hold the majority of project financing debt internationally. However, prior to the financial crisis in 2008 the industry was faced with an uncertain global economy and low interest rates. In response to the financial crisis, stricter regulations can into force, enforced by such regulatory instruments as the Basel III and Dodd-Frank Act, which did not allow banks to take on high risk loans and encouraged an increase in capital held (KPMG, 2016). These new regulations promoted short-term, low-risk loans which increased the difficulty and expense for securing long-term loans such as those used to fund infrastructure projects.

Therefore, attracting investors requires the right economic conditions. Conditions such as the current low interest rate environment has resulted in investors seeking alternative asset investment opportunities. And although infrastructure is considered an alternative investment opportunity that could realise increased profits, attracting the private investor is difficult for a myriad of reasons. Bielenberg et al. (2016) summarise these as:

1 Lack of transparency and bankable pipelines;
2 High-rate transactions and increase in prices;
3 Absence of a feasible funding structure;
4 Inadequate risk-adjusted returns;
5 Ambiguity; and
6 A lack of suitable regulations and policies.

A study completed by the OECD in 2018 found that commercial banks were constrained by several factors that prevented them from investing in infrastructure. The key factors inhibiting investment included:

1 Drawbacks of credit quality;
2 The up-scaling of competition in investment market mutual funds;
3 The instruction conditions of minimum risk-based capital demands;
4 The competitiveness of investment banks, namely in structure and syndicate loans; and
5 Restricted procedures in the aftermath of the GFC-2008, namely the re-scheduling of debts.

Private equity is another widely used source of private finance and funding for infrastructure projects which entails private investment in companies that are not publicly traded. As private equity funds tend to invest in companies that are not publicly traded, the risk of return is increased and therefore return can be inflated or deflated through the operational lifetime of the investment which can cause concern for some investors (Gemson et al., 2011). The scale and quality of the infrastructure project and its location can be a key driver for attracting private equity investment. The Australian Government have designed various venture capital programmes as an attraction mechanism for foreign and domestic equity investment;thus allowing businesses to develop technologies and convert research to product and encourage business growth (GOV. AU, 2021). As their aim is to exit at a higher equity than at which they entered, this growth is encouraged by equity investors. Between entry and exit is usually a period of 5–7 years. This is a relatively short period of investment when considering infrastructure projects and therefore, generally, may not be a suitable funding and financing mechanism.

The infrastructure deficit has had greater impact due the outbreak of the Coronavirus which has weakened economies across the world. Consequently, an infrastructure investment plan must be developed to tackle economic depression. Currently, the pandemic has had a vast impact on global economies with GDP levels in the UK 6.3 per cent lower than what they were in February 2020 at the time of the outbreak. The coronavirus has sent a shock of uncertainty through global markets creating an ambiguous environment for infrastructure investors (ONS, 2021).

According to Cohen (2016), public and private sector investment are both needed if the UK's infrastructure requirements are to be met: 'The crucial question is not whether to use private finance, but rather how it can be involved most efficiently'. Efficient use of private investment is therefore a challenge and barrier to investment. Governments must acknowledge that this barrier will remain due to the complexity and multiple stakeholder interest.

The NCE (Floater et al., 2017) global review identified seven key finance mechanisms that could have significant potential for overcoming barriers to investing in sustainable infrastructure.

1. Fiscal decentralisation
2. Bonds and debt financing
3. Land value capture
4. Pricing, regulation and standards
5. National investment vehicles
6. International finance
7. Public–private partnerships.

Inherent governance challenges and barriers must be overcome to attract private investment in infrastructure and realise a sustainable investment in infrastructure.

Understanding the political barriers to infrastructure investment

Political barriers to investment are essentially an inherently political choice, which potentially can stimulate or blunt economic growth, limit social inclusion, and evade environmental protections. Predicting how much investment is needed is complex but several organisations have made progress in this area, developing analytical tools and commentary, albeit using somewhat different approaches, to compare spending requirements over many nations to determine future investment needs. To deliver more investment in essential infrastructure, global leaders must acknowledge the present barriers to investment and develop strategies and mechanisms to overcome them. At the governmental level, finance ministers often lack infrastructure expertise, tending to view the topic as 'too technical'. There is therefore a need to translate the language of infrastructure into a fiscal and a macro business case. On a positive note, however, municipals are increasing in stature due the understanding that resilient local communities are fundamental in sustaining regional economies.

Therefore, governments must build investment-friendly systems, at both national and regional levels, that can instil confidence and encouragement in the private sector. To assemble such systems will require changes to financial and regulatory processes, built on solid political support for infrastructure. Ultimately government policy will determine the extent of private investment in infrastructure. This may be initiated through the purchase of a whole national industry, for example, the rail network or water (Howes and Robinson, 2005). In addition, because of the scale of infrastructure projects, and the risk associated with long-term investment, it is necessary that the financial feasibility of a project is fully investigated from the start. And, moreover, for private sector investment, which can come in many forms, governments need to ensure there is an adequate return, i.e. between 8–10 per cent.

Many infrastructure projects are, however, not appropriate for private sector investment initiatives because of the long-term nature of the projects and their associated risks. Add to this the decline in lending from commercial banks for such projects since the financial crisis, it means that there is an international shortage of funding. According to Bhattacharya et al. (2019) the G20 has become a 'champion for scaling up infrastructure investment and financing', and have called for the establishment of bankable projects, to attract the private sector and make more investment funding available.

The global economy, political uncertainty, and social conflict continue to have major impacts on national government revenues, ergo infrastructure investment. Since the global fiscal crisis governments' attitudes to PPP, particularly PFI (Private Finance Initiative), have changed. Before the crisis there was widespread political support for PFI. However, with the financial hardship that followed, government spending was scrutinised, resulting in a political backlash on PFI and on the role of private sector financial institutions in delivering and investing in national infrastructure. Criticism from politicians that PFI contracts were too expensive and inflexible ultimately led to the scrapping of PFI as a method of procurement for new public infrastructure as discussed previously.

Hence, a lack of political will has stalled infrastructure investments. However, in recent years governments have sought to better understand the value of the private sector in delivering infrastructure. It is no longer accepted that PPP should be used to move a large amount of government-related debt off the government's balance sheet. PPP has now become a collaboration between the public and private sectors insofar that the public sector financially contributes to the project and engages as a key stakeholder and decision maker. Yet negativity around PPP persists, with existing PFI projects plaguing governments who are increasingly seeking ways of renegotiating, or even exiting, existing PFI contracts. Ultimately government may believe that a project is in the national interest, and may even prefer it to be publicly funded, but the required level of funds may simply not be feasible within its fiscal strategy.

Yet, Howes and Robinson (2005) note that not all infrastructure projects are suited for PPP as the lack of revenue generated along with the risks involved outweighs the benefits. PPPs are significantly complex, demanding more highly specialised resources and attention by the government and do not suit those governments that need quick results (APMG International, 2019).

In developing economies political instability is considered to be the main obstacle encountered by private investors and consequently reduces their likelihood of investing. (Alderighi et al., 2019). Key factors that make investing in developing economies difficult are social unrest, economic instability, political unrest, and lack of knowledge and experience. In addition, corruption represents a further political challenge, leading to increased risk when investing in emerging countries (Loxley, 2013). Ironically the need for infrastructure is greatest in countries with the lowest incomes, yet they face the greatest barriers in accessing finance to aid the implementation of new systems and assets.

It is clear that government will be responsible for most of infrastructure investment and will have an influential role attracting private investment. Therefore, governments must also ensure that infrastructure investments are also meeting the agreed COP26 Glasgow Climate pact of delivering net zero emission by 2050. Mitigating and adapting to climate change and improving resilience and overall sustainability of infrastructure investment is now a priority. Governments are obligated to de-carbon infrastructure so as to reduce the impacts of climate change. Heatwaves were reported by the World Meteorological Organization (2019) to be the deadliest meteorological hazard, while tropical cyclones were associated with the largest economic losses, attributed to flooding and its associated loss and damage. Hurricane Harvey in 2017 is reported to have caused an estimated economic loss of more than US$125 billion. Hence, the greening of infrastructure at scale is a precondition for achieving sustainable growth (WEF, 2013) (see Figure 6.1).

Nowadays, any investment in infrastructure must be sustainable investment which will necessitate investments across multiple sectors supported by new green business models and green finance structures. According to the Global Infrastructure Hub (2021), environmental and social governance factors are of increasing importance for private investors looking to manage and mitigate risk and enhance financial performance and returns from infrastructure investments. In the USA green private investment in infrastructure projects has been increasing since 2014, rising from US$58 billion in 2014 to US$87 billion in 2020. However,

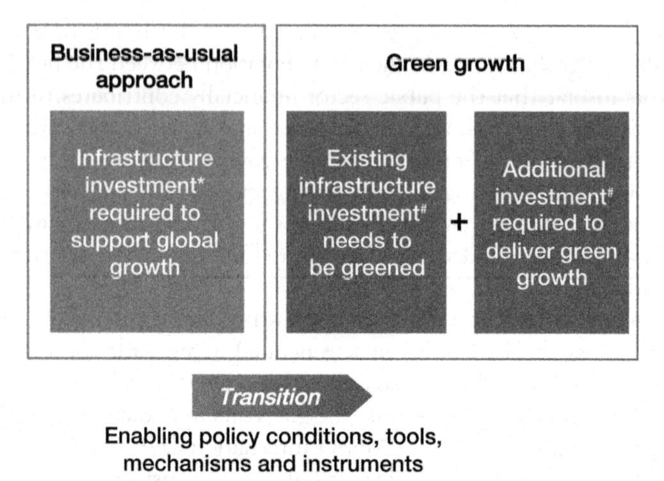

Figure 6.1 The transition to green infrastructure and sustainable growth
Source: WEF (2013).

greater efforts are required to reduce carbon emissions from infrastructure development across all infrastructure sectors.

Although the Global Infrastructure Hub (2021) considers this focus on renewables to be encouraging, it suggests that at its current level it is not sufficient to reach net zero targets. In fact, it states that wind and solar capacity additions must quadruple by 2030 to reach global net zero emissions by mid-century. Currently green private investment is reflected most in the renewables sector; other sectors need to make changes to increase their green investment.

Overcoming the key barriers to infrastructure investment

To develop 'best practice', there is a need for a joined up approach to financing infrastructure and the sharing of ideas. Howes and Robinson (2005) call for an alignment between strategic economic planning and the implementation strategy for projects, which in turn will allow for an appropriate balance between public and private funding. There is also a recognition that the burden of restarting the economy cannot rest entirely with government. Private businesses can invest in infrastructure and in doing so create jobs, apprenticeships, reskilling and education to the benefit of society. Infrastructure is often overlooked by the financial markets and investors because it is not viewed as a traditional asset class. The momentum generated by the Build Back Better UK Government initiative (HM Treasury, 2021) provides an opportunity to rejuvenate and update the traditional PPP model. The UK's recent Construction Playbook (HM Government, 2020), which sets out expectations for contracting parties, is an example of how governments are looking to improve infrastructure delivery. Elsewhere, in Canada, the PPP Alliance Model, built on the principle of 'mutual gain and pain' has been developed and deployed which signifies the Government's ability to become an agile private sector business partner.

The World Bank suggests three institutional pillars are needed to raise the probability of PPP success: Political will, institutions and governance, and project times (see Figure 6.2). Inderst (2020) gives the following general policy recommendations for countries, both developed and developing, wishing to catalyse institutional capital for infrastructure investment and finance:

1. Consistent infrastructure policies with a clear, stable regulatory framework and good public governance are essential for 'quality infrastructure' (a G20 concept).
2. No retrospective changes of rules and regulations; PPPs especially require time and a high degree of trust to succeed.
3. Strengthen the public sector capabilities not only in central government but also at the important sub-national levels (where it is most needed, especially in social sectors).
4. National infrastructure plans to include also social infrastructure, or set out separately.
5. Enlarge and enhance the pipeline of investable (social) infrastructure projects.
6. Consider 'asset recycling' (i.e. privatisation of operational assets, using proceeds for new, initially more risky or 'difficult' social facilities). 'Value capture' is one mechanism for the public sector to regain some of the indirect benefits of projects.
7. Creation of a public–private EU fund for social infrastructure, and recommendations for wider regional support policies.

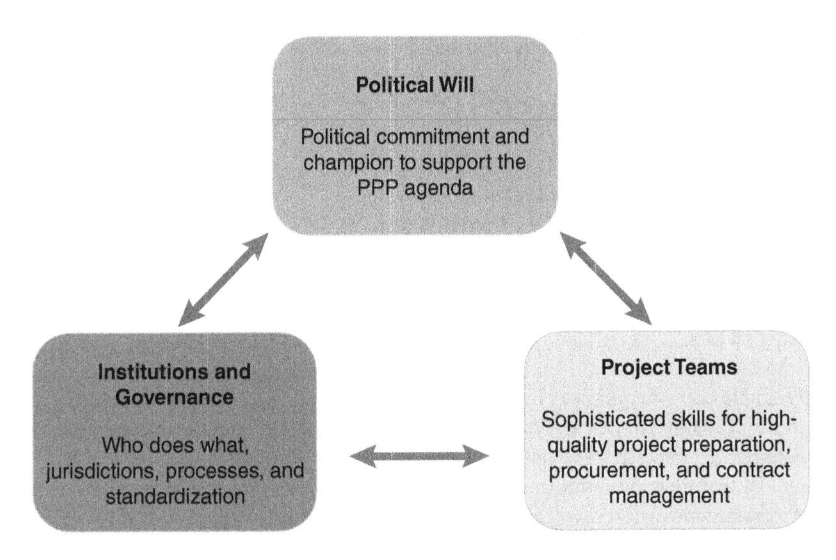

Figure 6.2 Three institutional pillars are needed to increase the probability of PPP success
Source: World Bank (Flor, 2018).

Summary

At this time of unprecedented fiscal distress, the world's most powerful economies must put infrastructure at the heart of their inclusive, sustainable and resilient recovery plans.

Modern and sustainable infrastructure is the key pillar for economic, social and environmental prosperity, and investment in infrastructure is important for increasing quality of life. Any government should consider investment in infrastructure as the core strategy for overall development.

Governments must develop appropriate goals and objectives for the development of infrastructure. Encouraging private sectors along with the public sector will boost the growth regional economies.

However, there are many challenges to infrastructure investment, and governments must grapple with the multi-dimensional aspect of doing more across government obligations with diminishing financial resources. The investment–growth nexus in the post-pandemic era will both expose the historic underspend within the sector and drive the efficiency and effective planning and delivery agenda of sustainable infrastructure at national and city level. Inherent governance challenges and barriers must be overcome to achieve a successful recovery towards the elusive sustainable infrastructure.

References

Alderighi, S., Cleary, S. and Varanasi, P. (2019). "Do institutional factors influence cross-border portfolio equity flows? New evidence from emerging markets", *Journal of International Money and Finance*, Issue 99, (2019)102070

APMG International (2019). "Disadvantages and pitfalls of the PPP option". Available at: https://ppp-certification.com/ppp-certification-guide/54-disadvantages-and-pitfalls-ppp-option

ASCE (2021). "ASCE Statement on GOP Infrastructure Framework". 22 April, American Society of Civil Engineers, Reston, VA. Available at: https://www.asce.org/publications-and-news/civil-engineering-source/society-news/article/2021/04/22/asce-statement-on-gop-infrastructure-framework

Bhattacharya, A., Gallagher, K.P., Muñoz Cabré, M., Jeong, M. and Ma, X. (2019, June). *Aligning G20 Infrastructure Investment with Climate Goals and the 2030 Agenda*. Foundations 20 Platform, a report to the G20. Available at: https://www.foundations-20.org/wp-content/uploads/2019/06/F20-report-to-the-G20-2019_Infrastrucutre-Investment.pdf

Bielenberg, M.K., Oppenheim, J. and Roberts, M. (2016, January). *Financing Change: How to Mobilize Private-Sector Finance for Sustainable Infrastructure*. McKinsey & Company. Available at: https://newclimateeconomy.report/workingpapers/wp-content/uploads/sites/5/2016/04/Financing_change_How_to_mobilize_private-sector_financing_for_sustainable-_infrastructure.pdf

Bitsch, F., Buchner, A. and Kaserer, C. (2010). "Risk, return and cash flow characteristics of infrastructure fund investments", *EIB Papers*, vol. 15, no. 1, pp. 106–136.

Carroll, A. (2022). "The future of infrastructure: Digitalisation, decarbonisation and demographics". Infrastructure Investor, 1 April. Available at: https://www.infrastructureinvestor.com/the-future-of-infrastructure-digitalisation-decarbonisation-and-demographics/

CBI (2020). "Private sector can help deliver UK's infrastructure revolution". 7 September. Available at: https://www.cbi.org.uk/media-centre/articles/private-sector-can-help-deliver-uks-infrastructure-revolution

Coalition for Urban Transitions (2017). "Global review of finance for sustainable urban infrastructure". Available at: https://urbantransitions.global/en/publication/global-review-of-finance-for-sustainable-urban-infrastructure/

Cohen, G. (2016). "Financing the UK's infrastructure: Private and public gains". *The Economist*, 19 April. Available at: https://eiuperspectives.economist.com/economic-development/vibrant-economy/blog/financing-uks-infrastructure-private-and-public-gains

European Investment Bank (2021). *Global Solutions, International Partnerships. The European Investment Bank Development Report 2021.* Available at: https://www.eib.org/attachments/thematic/the_eib_development_report_2021_en.pdf

Floater, G. et al. (2017). 'Global review of finance for sustainable urban infrastructure'. Background paper, New Climate Economy, Coalition for Urban Transitions. Available at: http://newclimateeconomy.report/workingpapers/wp-content/uploads/sites/5/2018/01/NCE2017_CUT_GlobalReview_02012018.pdf

Flor, L. (2018). "Three ways governments can create the conditions for successful PPPs". World Bank Blogs. Available at: https://blogs.worldbank.org/transport/three-ways-governments-can-create-conditions-successful-ppps

Gemson et al. (2011). "Impact of private equity investments in infrastructure projects". Available at: https://www.sciencedirect.com/science/article/abs/pii/S095717871100083X

Global Infrastructure Hub (2021). "Infrastructure monitor". Available at: https://www.gihub.org/infrastructure-monitor/and https://cdn.gihub.org/umbraco/media/4410/gihub_v10.pdf

GOV.AU (2021). "Venture capital". Available at: https://business.gov.au/grants-and-programs/venture-capital

Haider, Z. (2017). *Parallel Perspectives on the Global Economic Order: A U.S.–China Essay Collection Report*, Published by Center for Strategic and International Studies (CSIS) pp. 84.

HM Government (2020, December). *The Construction Playbook: Government Guidance on Sourcing and Contracting Public Works Projects and Programmes*. Available at: https://www.gov.uk/government/publications/the-construction-playbook

HM Treasury (2020, November). *National Infrastructure Strategy: Fairer, Faster, Greener*, CP 329 (London: HM Treasury). Available at: https://assets.publishing.service.gov.uk/government/uploads/system/uploads/attachment_data/file/938049/NIS_final_web_single_page.pdf

HM Treasury (2021, March). *Build Back Better: Our Plan For Growth*. CP 401. Available at: https://assets.publishing.service.gov.uk/government/uploads/system/uploads/attachment_data/file/968403/PfG_Final_Web_Accessible_Version.pdf

Howes, R. and Robinson, H. (2005). *Infrastructure for the Built Environment: Global Procurement Strategies*, Butterworth-Heinemann, Oxford.

Inderst, G. (2020). "Social infrastructure finance and institutional investors: A global perspective". Discussion Paper, ZBW – Leibniz Information Centre for Economics, Kiel, Hamburg. Available at: http://www.ltiia.org/wp-content/uploads/2015/12/Inderst_SocialInfrastructureFinanceAndInstitutionalInvestors_Sept20.pdf

Kemna, A. (2015). "The impact of regulation". Available at: https://www.mckinsey.com/business-functions/strategy-and-corporate-finance/our-insights/the-impact-of-regulation

KPMG (2016). "Banking and private capital: Friend or foe?". Available at: https://assets.kpmg/content/dam/kpmg/pdf/2016/05/ie-private-capital-friend-or-foe.pdf

Loxley, J. (2013). 'Are public–private partnerships (PPPs) the answer to Africa's infrastructure needs?'. *Review of African Political Economy*, vol. 40, no. 137, pp. 485–495.

McKinsey (2020). "Governments can get the most out of infrastructure projects". Available at: https://www.mckinsey.com/industries/public-and-social-sector/our-insights/four-ways-governments-can-get-the-most-out-of-their-infrastructure-projects

OECD (2015, September). *Risk and Return Characteristics of Infrastructure Investment in Low Income Countries*. Report, 3 September. Available at: https://www.oecd.org/pensions/private-pensions/Report-on-Risk-and-Return-Characteristics-of-Infrastructure-Investment-in-Low-Income-Countries.pdf

OECD (2018). *Multilateral Development Finance: Towards a New Pact on Multilateralism to Achieve the 2030 Agenda Together*, OECD Publishing, Paris.

ONS (2021). "Coronavirus and the impact on output in the UK economy: December 2020". Available at: https://www.ons.gov.uk/economy/grossdomesticproductgdp/articles/coronavirusandtheimpactonoutputintheukeconomy/december2020

Sherraden, S. (2011). "The infrastructure deficit policy paper". Available at: https://www.newamerica.org/economic-growth/policy-papers/theinfrastructure-deficit/

The World Bank (2020). "Investors in infrastructure in developing countries". Available at: https://ppp.worldbank.org/public-private-partnership/financing/investors-developing-countries

Woetzel, J., Garemo, N., Mischke, J., Kamra, P. and Palter, R. (2017, October). "Bridging infrastructure gaps: Has the world made progress?". Discussion Paper, McKinsey Global Institute, McKinsey & Co, USA. Available at: https://www.mckinsey.com/business-functions/operations/our-insights/bridging-infrastructure-gaps-has-the-world-made-progress

World Built Environment Forum (2020, October). "The economics of infrastructure and the post-COVID-19 recovery". WBEF, RICS, 28 October. Available at: https://www.rics.org/uk/wbef/megatrends/markets-geopolitics/the-economics-of-infrastructure-and-the-post-covid-19-recovery/

World Economic Forum (2013). *The Green Investment Report: The Ways and Means to Unlock Private Finance for Green Growth*. Available at: https://www3.weforum.org/docs/WEF_GreenInvestment_Report_2013.pdf

World Meteorological Association (2019). "2019 concludes a decade of exceptional global heat and high-impact weather". Press Release, 3 December. Available at: https://public.wmo.int/en/media/press-release/2019-concludes-decade-of-exceptional-global-heat-and-high-impact-weather#:~:text=2019%20is%20on%20course%20to,above%20the%20pre%2Dindustrial%20period

Index